# Image Fusion in Preclinical Applications

Claudia Kuntner-Hannes • York Haemisch
Editors

# Image Fusion in Preclinical Applications

Springer

*Editors*
Claudia Kuntner-Hannes
Center for Health & Bioresources
AIT Austrian Institute of Technology GmbH
Seibersdorf
Austria

York Haemisch
Technical Sales & Business Development
Direct Conversion GmbH
München
Germany

ISBN 978-3-030-02972-2 ISBN 978-3-030-02973-9 (eBook)
https://doi.org/10.1007/978-3-030-02973-9

Library of Congress Control Number: 2018964720



This Springer imprint is published by the registered company Springer Nature Switzerland AG
The registered company address is: Gewerbestrasse 11, 6330 Cham, Switzerland

# Preface

In the last decade, our understanding of the function of the human body and the interaction of drugs with certain molecular targets has increased significantly. This deepened insight is a result of the wider availability of new diagnostic tools such as those developed by the different "-omics" disciplines (genomics, proteomics, etc.), but largely also of the application of noninvasive, in vivo imaging techniques applying molecular targets and markers. In vivo imaging modalities have different characteristics based on their individual mechanisms of tissue contrast or function, specific sensitivity, and spatial and temporal resolution in relationship to diseases or biological processes. Multimodality imaging using two or more imaging modalities simultaneously or sequentially allows the combination of the strengths of individual modalities while overcoming their limitations. Anatomical imaging techniques such as (X-ray) computed tomography (CT) and functional imaging such as magnetic resonance imaging (MRI) provide excellent structural detail and resolution. Molecular imaging techniques such as positron emission tomography (PET) and single-photon emission computed tomography (SPECT) allow tracing of biochemical and biological processes at the molecular level. By combining anatomical and/or functional with molecular imaging, complementary information can be obtained leading to an improved understanding of the physiological mechanisms at the molecular and cellular level.

In vivo imaging modalities have emerged from their separate use to application in combination, generating multimodal datasets. The combination of multimodal datasets by image fusion has led to a new understanding of the biology underlying multiple diseases and biological processes. Multimodal imaging techniques can be based on the acquisition of images at different times, on the acquisition of images using different techniques, or on simultaneously acquired images. Especially in preclinical imaging, using animal (mostly rodent) models of human diseases, a variety of imaging tools are available, ranging from anatomical imaging with CT and anatomical and functional imaging with MRI to the molecular imaging techniques based on the use of isotopes such as PET and SPECT. Especially for imaging small rodents, these techniques are complemented by optical or acoustic imaging techniques such as optical imaging, ultrasound, and optoacoustic imaging. Using this enormous toolset of techniques allowing noninvasive insights into objects, imaging has the potential to improve the identification and development of new diagnostic and therapeutic drugs and to facilitate the translation of preclinical findings to applications in the clinics.

This book attempts to provide an overview of the different ingredients that are important in preclinical multimodal imaging such as image fusion techniques, data analysis, imaging systems hardware, and animal handling. Applications of various combinations of imaging techniques are explained exemplarily on different case studies in cardiovascular diseases, inflammatory diseases, radiotherapy, and drug development applications. The reader might find it particularly interesting to learn about the use of an imaging modality ranging from macroscopic (tomographic) imaging of whole animals down to microscopic imaging of cells using optoacoustics.

While this compilation of experiences of acknowledged authors in this field can only point some spotlights onto interesting aspects of preclinical in vivo imaging, we still hope that this book will enable the reader to obtain an understanding of the different factors and aspects to be considered when planning and performing preclinical imaging experiments. We also wish that the case studies provided might serve as examples of how to apply multimodal imaging techniques adapted to different research questions and challenges in many application areas and might trigger inspiration and ideas on when and how to apply multimodal preclinical imaging in your own research studies.

Seibersdorf, Austria — Claudia Kuntner-Hannes
München, Germany — York Haemisch

# Contents

# High-Level Story: Data Analysis in Multimodal Preclinical Imaging—Methods and Tools

# 1

Gabriel Tobon, Jacob Hesterman, Shil Patel, and Christian Lackas

## 1.1 Introduction

Preclinical research has long been at the forefront of software and methodological innovation for multimodal imaging. In vivo imaging, ex vivo imaging, and the combination of the two offer a wide variety of complementary image modalities for deriving biological insight. It is increasingly common for research applications to use images from hybrid modality scanners, images from multiple scanners, and derived images in tandem for advanced analysis and quantitation. A long-standing example of the value of multimodal imaging comes from positron emission tomography (PET) and single-photon emission computed tomography (SPECT) imaging, where it is a common practice to acquire a computed tomography (CT) image for attenuation correction (AC) of the reconstructed signal [1]. It demonstrates how combining image modalities can not only provide complementary information but also enhance the quality and reliability of one or more of the modalities. In the subsequent analysis of PET and SPECT images, it is common for a researcher to quantify the sum or concentration of signal within a region of interest. As "functional" modalities, they may only exhibit contrast to background in regions where the targeted function takes place; however, more regions than those visible may be desired a priori for quantification. The quantitation of such regions greatly benefits from fusing an anatomical modality with a functional one for use in region segmentation.

To maximize the insight derived from multimodal applications, careful alignment of images and their corresponding fusion is required for visualization. It is

G. Tobon · J. Hesterman · C. Lackas (✉)
Invicro—A Konica Minolta Company, Boston, MA, USA
e-mail: lackas@invicro.com

S. Patel
Eisai Inc., Andover, MA, USA





C. Kuntner-Hannes, Y. Haemisch (eds.), *Image Fusion in Preclinical Applications*, https://doi.org/10.1007/978-3-030-02973-9_1

software's function to assist the user in establishing the spatial correspondence across images and present their combined information appropriately. In the simplest case, hybrid scanner vendors provide hardware-level alignment parameters, a fixed transform that can be used by the software to align images for fusion. Otherwise, if modalities are acquired on separate scanners, an inherent balance exists between images containing complementary information for added insight and shared information to support adequate alignment. Sophisticated alignment methods using mutual information-based similarity metrics support the co-registration with minimal shared spatial patterns. In yet another level of difficulty, it may be necessary to have pixel-wise alignment across subjects in a multi-subject study. Functional magnetic resonance imaging (fMRI) and atlas-based approaches often require pixel-wise spatial correspondence across subjects to normalize spatial patterns to a template image matrix. In other cases, precise pixel-wise alignment may not be necessary, yet template co-registration can more clearly normalize structure when comparing subjects visually. Indeed, significant efforts have been made to combine multimodal images acquired across patients, species, and imaging sites to improve understanding and translation in research studies [2]. A truly multimodal software package should be able to load and fuse image data stored in varied formats, matrix sizes, voxel information, units, and dynamic ranges and leverage a variety of co-registration algorithms.

This chapter is divided into three sections. In the first section, popular software options used for preclinical research, core functions provided by these software packages for image fusion, and the application of these tools on recent preclinical studies are reviewed. In the second section, methods and examples are provided within which multimodal data are utilized for the extraction of quantitative data. This section focuses especially on how image registration tools, such as those available within the software platforms described in the first section, enable application of anatomical data to improve estimations from functional data. The final section provides examples of how multimodal data, often across a variety of spatial or temporal scales, can provide a more comprehensive understanding of a biological question. This section focuses particularly on how software tools are used to help combine and visualize such data. Finally, samples of novel analysis and visualization routines used to combine multimodal data derived from recent research in machine learning and medical image processing are introduced and looked at.

## 1.2 Software Options

Although many clinical viewing software packages can be repurposed for preclinical imaging applications, preclinical image data introduce a varied set of formats and unique challenges not commonly supported by clinical imaging software tools. In preclinical research, the focus on diagnostic accuracy is supplanted with issues of spatial resolution and noise reduction. In addition, image data formats vary in their levels of compliance to the DICOM standard and can often be proprietary. Table 1.1

**Table 1.1** A list of commonly used software in preclinical radiology research

| Software | Author | Licensing | Application area | Platforms |
|---|---|---|---|---|
| VivoQuant[a] [3] | Invicro | Commercial | General | Windows, Mac OS, GNU/Linux |
| ImageJ/Fiji [4] | Community | Free (BSD/GPL) | General | Windows, Mac OS, GNU/Linux |
| SPM [5] | UCL | Free (GPL) | MRI, PET/SPECT, EEG/MEG | Windows, Mac OS, GNU/Linux on MATLAB |
| FSL [6] | Oxford | Free (FSL) | MRI | Mac OS, GNU/Linux |
| Amide [7] | Stanford | Free (GPL) | General | Windows, Mac OS, GNU/Linux |
| Analyze [8] | AnalyzeDirect | Commercial | General | Windows, Mac OS, GNU/Linux |
| AFNI/SUMA [9] | NIH | Free (GPL) | MRI | Windows (Cygwin), Mac OS, GNU/Linux |
| Paraview [10] | Kitware[b] | Free (BSD) | General | Windows, Mac OS, GNU/Linux |
| ITK SNAP [11] | UPenn/Utah/Kitware | Free (GPL) | General | Windows, Mac OS, GNU/Linux |
| PMOD [12] | PMOD Technologies | Commercial | PET/SPECT, CT, MRI | Windows, Mac OS, GNU/Linux |
| 3D Slicer [13] | Community | Free (Slicer) | General | Windows, Mac OS, GNU/Linux |
| FreeSurfer [14] | MGH Martinos Center | Free (Freesurfer) | MRI, PET | Mac OS, GNU/Linux |
| Amira [15] | FEI | Commercial | General | Windows, Mac OS, GNU/Linux |
| Osirix [16] | Pixmeo | Commercial (Free Lite version) | General | Mac OS |

The authors listed are not comprehensive; however, they indicate the primary development contributors to the application. Free licenses are marked by name and range from copyleft, to free for academic use, and flexible BSD-style varieties. Application areas and supported platforms are marked according to how the software product is advertised; however, it is common that different software packages specialize or lack in methodologies not indicated in this table

[a]VivoQuant was used in the generation of all figures used in this chapter

[b]ParaView contributors also include Sandia National Laboratories, Los Alamos National Laboratory, Army Research Laboratory, and CSimSoft

catalogs some of the most prevalent imaging software tools used by preclinical researchers.

Each software specializes in certain analysis and visualization applications, and users must consider data format compatibility between their imaging equipment and the analysis software. Many camera vendors will provide a custom-developed software not mentioned in Table 1.1 for the visualization and analysis of images from their corresponding instrument; however, these tools often lack capabilities required for multimodal multi-scanner analysis applications. Examples of such software include ParaVision® from Bruker and Living Image® from PerkinElmer. Users also have a choice between commercial and free applications. Free applications with open-source licensing often receive novel contributions from the research community. By exposing their source code, they provide users with the deepest level of flexibility and customization, thus making them a strong choice for method development researchers. By contrast, commercial applications tend to hide their algorithmic complexity in an effort to make workflows more accessible to general users and less susceptible to user errors, and their development teams are more likely to invest in government regulatory compliance. In this chapter, all example images and analyses are generated by VivoQuant®, an application specializing in multimodal visualization and analysis for CT, PET, SPECT, MRI, ultrasound, autoradiography, and fluorescence imaging with the ability to load a variety of open and proprietary preclinical imaging formats.

Much of the software in Table 1.1 relies on core technology developed by contributors to open-source and proprietary medical imaging libraries. These libraries include, but are not limited to, the Insight Toolkit [17] for segmentation and co-registration, Visualization Toolkit [18] and Open Inventor [19] for rendering, Qt [20] for user interfaces and networking, DCMTK for DICOM support, and Java [21] for cross-platform deployment. Indeed, many software developers in the imaging community both collaborate and compete to advance the tools for imaging scientists. To be effective for multimodal applications, tools should support a variety of image formats, fuse images together with sophisticated co-registration and visualization algorithms, and provide both generalized and domain-specific processing tools.

## 1.3 Extraction of Data for Quantitation

The ability to extract estimates of physical or physiological quantities is a powerful feature of imaging. Quantitative accuracy depends on many factors, such as scanner calibration, noise, spatial and temporal resolutions, quality of segmentation, and application of appropriate corrections and models. Image registration, both intra- and inter-modality, plays a critical role in accurate quantitation. This importance is often manifested through the use of image registration to enable the generation and/or application of anatomical regions of interest (ROIs).

### 1.3.1 Preprocessing

Most image processing pipelines include some form of preprocessing or data harmonization, followed by a ROI generation procedure and extraction of quantitative data from the generated ROIs. In some cases, those extracted data undergo additional processing (e.g., tracer kinetic modeling). While several data processing steps (e.g., intensity normalization, resampling, unit conversion) are unimodal in nature, multimodal data are critical in many applications to improve the accuracy and robustness of this data extraction process.

The preprocessing step for which multimodal data are most critical is that of image alignment or co-registration. In some multimodal scanners acquiring images simultaneously, different modalities are natively co-registered via its reconstruction algorithm. Commonly, however, modalities must be acquired in separate scanning sessions, perhaps on different scanners entirely, during which the patient/animal has been moved or adjusted. In such cases, a post hoc co-registration is needed to align images prior to further analysis. In the case of preclinical imaging, it is not uncommon for researchers to use rodent "hotels" where multiple animals can be scanned simultaneously in one session. These images must be cropped and aligned appropriately across modalities to analyze animals individually. Figure 1.1 shows an example where the PET image and a CT image of a mouse hotel were acquired on separate scanners. The individual animals were segmented using the CT scan, and those segmentations were applied to crop both the PET and CT images. This is a

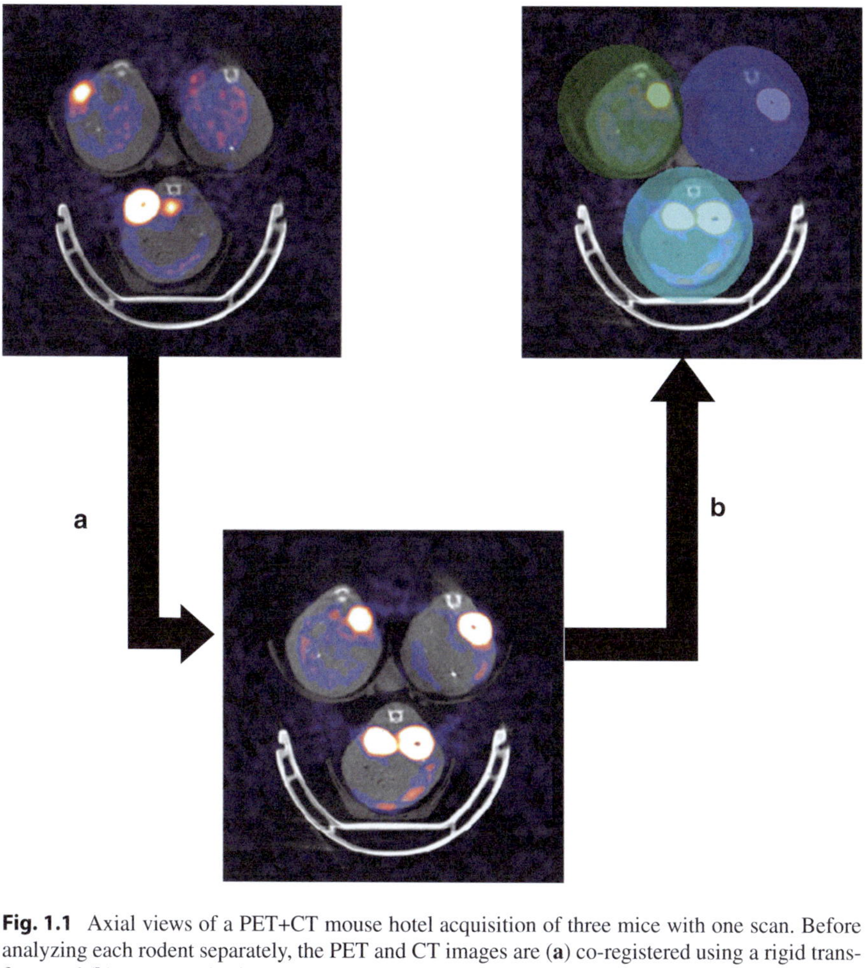


**Fig. 1.1** Axial views of a PET+CT mouse hotel acquisition of three mice with one scan. Before analyzing each rodent separately, the PET and CT images are (**a**) co-registered using a rigid transform and (**b**) segmented using an automatic laplacian-based algorithm

canonical approach for multimodal software—derive an image transformation from one modality and apply it to one or more additional modalities.

A co-registration between two images is typically defined with a spatial transformation applied to one image, the "moving" image, to align it more closely to the other, the "fixed" image. To visualize or analyze an aligned image, the moving image must be resampled spatially using the co-registration transform.

#### 1.3.1.1 Transforms

##### Linear

Linear spatial transformations can be subcategorized into rigid and affine transforms based on their mathematical guarantees. Rigid transforms preserve *absolute* distances between any two points of the image. A rigid transform parameterized by rotation and translation is often the best choice for multiple scans of one animal with stable structural anatomy, such as multiple scans of the brain. Thanks to the equivalence of absolute distances, it is expected that volumes, angles, and pixel quantitation will all be stable following the application of a rigid transform. Indeed, in practice, these values are stable for many naturally occurring structures within some sampling error inherent to the discrete nature of image data. More details on the effects of sampling error can be found in Image Registration Pitfalls.

Affine transforms guarantee the preservation of *relative* distances within an image and define a superset of rigid transforms. An affine transform may include parameters for scaling or shearing an image and thus loses many of the aforementioned guarantees provided by rigid transforms.

##### Nonlinear

In many imaging research applications, a linear transform will not be sufficient to provide adequate alignment. Group-level voxel-wise analysis and atlas-based volume estimation are commonly used cases for nonlinear spatial alignment, where similar modalities are co-registered across different patients [22–24] . This is often called "spatial normalization" because it normalizes scans from different animals with respect to their anatomical differences. Certain MRI routines, such as echo-planar imaging, introduce significant spatial distortion that can be partially recovered from using multimodal nonlinear alignment [25] or even alignment of serial acquisitions taken in opposite phase-encoding directions [26]. These artifacts can appear more severely in preclinical MRI than in clinical applications [27], further justifying the use of nonlinear spatial alignment. The potential gains possible from proper utilization of nonlinear registration are illustrated in the example shown in Fig. 1.2.

Various state-of-the-art nonlinear co-registration transforms and optimization algorithms exist for accurate alignment [28]. These transforms are often high dimensional and provide minimal guarantees for quantification. In most cases, the transform itself is spatially regularized during optimization to both avoid erratic transforms and preserve locally similar relative distances. Within a small enough local window of the image, the transform may be approximated linearly where

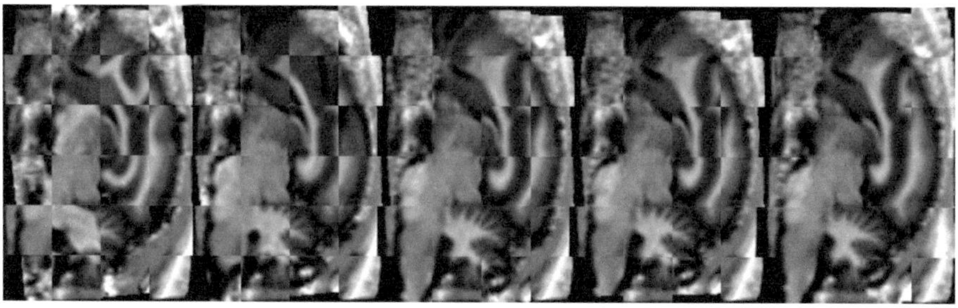

**Fig. 1.2** Example co-registration of T1 MRI maps of feline brains. From left to right, a checkboard overlay view of (*1*) two brains from different feline patients prior to co-registration, (*2*) the brains aligned with a translation transform, (*3*) the brains rigidly aligned, (*4*) the brains aligned with an affine transform, and (*5*) the brains aligned with a nonlinear deformation field computed using the symmetric normalization algorithm

relative changes in size are stable. Accurate transform estimation often requires that the images already be linearly aligned and that they contain dense anatomical contrast information. Indeed, nonlinear transforms are most accurately estimated from modalities with significant intensity heterogeneity, such as MRI and white light photography, although they have been successfully used with other modalities, such as CT and ultrasound [29, 30].

#### 1.3.1.2 Similarity Metrics

Co-registration algorithms are effective implementations of cost function optimization. The cost function to be minimized is typically a negated similarity metric. For example, the similarity between two images may be defined as the summed Euclidean distance between pairs of landmarks on the image defined by the user. Pinpointing consistent landmarks can be a challenging and manual task without automatically identifiable multimodal fiducials, so often a more global approach is used. An efficient landmark-based approach will often have consistent fiducials visible in both modalities and images to be co-registered as in Fig. 1.3.

For the following metrics, we first define two images—fixed image $I_f$ and moving image $I_m$—with intensity values $I_f(x)$ or $I_m(x)$ where $x$ is the $n$-dimensional spatial location of a voxel. The normalized correlation $\rho$ computes the spatial intensity correlation between images:

$$\rho = \frac{\mathrm{Cov}(I_m, I_f)}{\mathrm{Std}(I_m) \times \mathrm{Std}(I_f)}$$

where Cov is the sample spatial covariance function and Std is the sample spatial standard deviation function. This similarity measure shares the same advantages and pitfalls it would with two one-dimensional functions. The metric works well with similar image histograms, that is, similar modality and acquisition details. Without prior histogram equalization, a pair of T1- and T2-weighted images will not co-register well using normalized correlation.

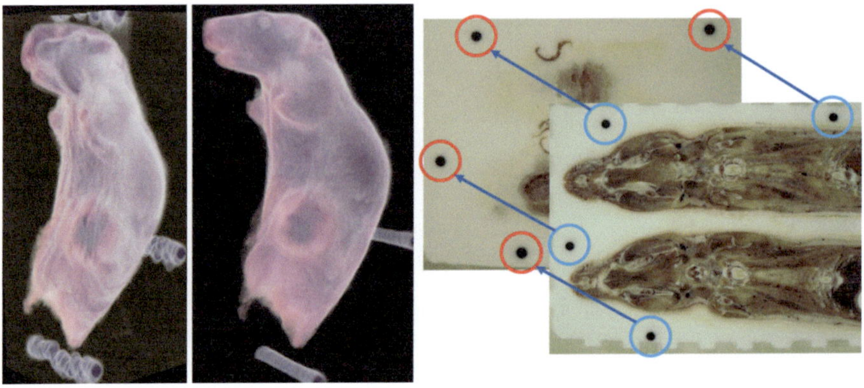

**Fig. 1.3** Fiducials suitable for landmark-based co-registration of ex vivo images. On the left are 3D reconstructions of white light images and fluorescence images, each with fiducials made through the acquisition block and used for 3D alignment. On the right, similarly embedded fiducials are matched across different slice photographs from a block. Co-registration of these images is done using the Euclidean distance between fiducial markers

Another commonly used metric is mutual information. The motivation behind using mutual information as a similarity metric is statistical: it measures how readily the fixed image's voxel intensities can be predicted from the moving image. This metric's computation begins by estimating the spatial probability density function (PDF) of intensities for each image. These PDFs are used to compute the statistical mutual information measure between the two images. Mutual information is better suited for multimodal images and heterogeneous acquisition parameters for co-registration. It is the default metric for co-registrations done in VivoQuant.

#### 1.3.1.3 Image Registration Pitfalls

The above paragraphs have outlined the utility of registration transforms in introducing efficiency and consistency gains to quantitation. However, because images exist on discrete grids, information content is not perfectly preserved upon application of a transform.

##### MR-Based Susceptibility Artifact Correction

Functional imaging data often refers to the nuclear modalities of PET and SPECT. However, several MRI modalities may also provide functional information. These functional MRI modalities often require specialized sequences which sacrifice the resolution and contrast of anatomical data to capture the functional data of interest. As an example, consider diffusion-weighted imaging (DWI). These scans often suffer from susceptibility artifacts at tissue boundaries which manifest themselves in image distortion. This distortion is limited to the phase-encoding direction. Using specific image registration steps, both linear and nonlinear, but specifically restricted to the phase-encoding direction, enables correction of susceptibility-induced distortion in the DWI image through the use of the anatomical information in a T1-weighted scan.

**Application of Image Registration for Quantitative Extraction**

We can define a region of interest in the moving image with volume $V_m$ and total intensity $I_m$ and thus a concentration:

$$C_m = \frac{I_m}{V_m}$$

Given a transform $T(\cdot)$, the software aligns the moving image into the fixed image space to produce new values for concentration $C_f$, total intensity $I_f$, and volume $V_f$. For a rigid transform, $V_f = V_m$, $I_f = I_m$, and $C_f = C_m$ within resampling error. An affine transform, however, may scale the region such that $V_f \neq V_m$; given the above relationship, it is clear to see that either $C_f = C_m$ or $I_f = I_m$ in this case, but not both. Expressed in other terms, either the *mean* intensity value or the *total* intensity value will be stable in the transformed region. It is important to have clarity over which has been preserved when an affine transform has been applied or avoid the potential confusion altogether by using a rigid transform.

The linearity property of rigid and affine transforms does afford the user accurate *relative* quantification. $V_f = |T| \times V_m$; thus anatomical structures change with a constant multiplicative factor across the entire image. This constant factor drops out of the equation when normalizing a region of interest with a reference region. This can be demonstrated by comparing the behavior of SUV and $SUV_R$ after nonrigid affine transformation. A standardized uptake value (SUV) is equal to the concentration within a volume (pixel or region) normalized by the injected dose per body mass. $SUV_R$ is the ratio between SUV within a tracer's target region and the SUV within a chosen nontarget, or "reference," region. A unit of voxel intensity such as SUV would require qualification if the image underwent affine co-registration but not rigid, while $SUV_R$ would be stable with any linear transformation.

Post-reconstruction, care must be taken in the subsequent analysis software to maintain the fidelity of quantitative data [31]. Imaging software users expect total uptake in a region of interest to stay constant after resampling. Although this is not generally true, many practical applications maintain equivalent region uptake and concentration values within a small amount of unbiased error. The magnitude of this error is controlled in many applications by the size and shape of the region itself. For a spherical region with radius $r$, it is expected that errors in the uptake will occur most at the surface of the sphere. The error is bound by the quantity of voxels at the surface and is proportional to $r^2$ compared to a total quantity of voxels proportional to $r^3$. The case for resampling is also aided by the fact that images are generally spatially smooth; thus the mixing of voxels inside and outside the region would not dramatically change local uptake values. Countermeasures can be taken to minimize the effect of resampling on quantitative analysis, including the use of 3D vs. 2D regions of interest, larger regions, partial voxel quantitation, fixed voxel sizes across data sets, and fixed volume regions of interest. Increasingly popular is the use of partial volume correction [32], which should be done prior to any needed resampling to avoid the introduction of errors into the procedure.

**PET/SPECT Dopamine Example**

Consider estimation of signal in the striatum of nuclear medicine (NM) images such as PET or SPECT acquired using a dopamine tracer. A test image is co-registered to an NM template, which may be either a single reference NM image or a composite reference (see Atlases). The ROIs, striatal regions in this case, will have either been generated on this NM template or manually co-registered to it. With each test data set co-registered to the NM template, this common set of ROIs may be applied to extract summary statistics. Variability is reduced through the use of these common ROIs, and visualization is aided by the mapping of all data to a common space.

There are potential issues with this approach. For example, consider normal biological variability in structure. Nonlinear registration transforms are required to capture this variability in mapping to the template space. However, the low resolution and lack of anatomical structure in many functional modalities prohibit successful application of nonlinear registration methods. Additionally, consider the situation of pathology, often of interest, has resulted in diminished signal in the regions of interest. While stiffness of elastic solvers may be tuned to prevent overfitting, this step is not typically sufficient to allow optimal registration to a template space using functional image data alone.

Therefore, multimodal registration techniques are critical. Structural imaging modalities, particularly T1-weighted or T2-weighted MRI, provide superior resolution and tissue contrast to those of functional image data. Atlases are created and aligned to MRI-based templates. Functional data are acquired in conjunction with MRI data for a test subject and aligned via rigid registration, assuming the functional and MRI data are acquired closely enough in time that structural changes will not have occurred. The test MRI may then be nonlinearly registered to the MRI template, and this transformation may be applied to the functional data to map it to the template space. This approach has become the standard practice in many preclinical and clinical applications.

### 1.3.2 Atlases

Atlases are population-level representations of anatomical structures and, as such, are a powerful tool for generating quantitative output in specific regions of interest (ROIs). This section describes some methods utilized for atlas construction and application.

Atlas construction relies on development of a reference data set and generation of labels. One approach to generating the reference data set is via combination of multiple individual data sets drawn from a representative population of interest. These data sets are combined through linear and nonlinear image registration to produce a composite template image that is representative of the overall population.

An additional step of composing the template image in a well-defined coordinate space is often applied to better enable cross-study comparison and dissemination across multiple imaging centers. Anatomical labels are typically manually defined by one or more expert anatomists on the template image.

Multimodal/functional atlases are utilized to apply ROIs to data sets in cases where the ROIs are not necessarily well-represented by the signal present in the test image data. In other words, it may be the case where signal characteristics prohibit direct segmentation or identification of a particular ROI on a test data set so the ROI is defined based on registration of the test data set and a pre-existing atlas. Such atlases are often constructed with and applied to multimodal data.

Probabilistic atlases are generated similarly to anatomical atlases. However, atlas labels are not deterministic in the sense that region boundaries are clearly defined and region membership is binary (inside or outside). Rather, on construction, labels from multiple individual data sets are combined to provide a probability value for a given voxel to belong to a particular region. Probabilistic atlases have the advantage of accounting for partial volume effects but can be more challenging to interpret and require a separate atlas instance for each region.

Table 1.2 provides examples of anatomical, functional/multimodal, and probabilistic atlases for preclinical and clinical use. The list is not comprehensive but provides a sample of the types of atlas approaches utilized in preclinical research.

#### 1.3.2.1 Multi-atlas Segmentation

Multi-atlas segmentation (MAS), also known as label fusion, methods utilize image registration techniques to map a population of reference subjects and their associated subject-specific region segmentations to a test subject, producing a probability map that may be used to segment regions within the test subject [50]. The MAS approach requires some initial setup time for population of a reference library but is powerful as a general segmentation approach across species, modality, and disease area. The segmentation method benefits from a multi-resolution approach. A cascade of registration and bounding box operations are utilized to map reference library data sets to an input test data set. For example, in a whole-body image, this process is achieved through a whole-body affine registration, followed by axial subregion affine registration, and, finally, region-specific nonlinear registration. The resulting affine and nonlinear transforms are combined and applied to the reference library ROIs. Reference library data sets are sorted by registration performance. A predefined number of registered reference library ROIs are combined to generate an ROI probability map representing the probability that any given voxel belongs to the test data set region of interest. A deterministic ROI may be generated from the ROI probability map using a variety of methods such as thresholding or maximum likelihood approaches [50].

#### 1.3.2.2 Brain Region Segmentation Without Soft Tissue Contrast

The anatomical imaging modality accompanying many preclinical nuclear imaging scanners is x-ray CT rather than MRI. CT images have many uses but tend to have less soft tissue contrast than anatomical MRI data. This lack of soft tissue contrast

**Table 1.2** A list of atlases available to preclinical researchers, including the species, modality, and accessibility information

| Atlas | Species | Construction (voxel/surface, automatic/manual) | Availability |
|---|---|---|---|
| Digimouse [33] | Mouse | Voxel model built from co-registered CT, PET, and cryosection data | Free |
| MOBY [34] | Mouse | Based on MRI and MRM data and modeled with NURBS. Also has mouse skeleton atlas, cardiac, and respiratory functions | Licensing fee through Johns Hopkins University |
| Shan T2 MR [35] | Rat | Voxel model | email author: shanbc@ihep.ac.cn |
| Allen Brain Atlas [36] | Mouse, human | Coronal and sagittal mouse sections | Online tools |
| XCAT Phantom [37] | Human | NURBS surfaces, designed for 4D | Licensing fee through Johns Hopkins |
| Paxinos and Watson [38] | Rat brain | Stereotaxic coordinates | Book |
| INIA19 [39] | Rhesus macaque brain | Volume, from 100 T1 MRI scans from 19 animals | *Online for download*, Attribution License |
| MNI Macaque Atlas [40] | *Macaca fascicularis*, *Macaca mulatta* | Stereotaxic reference frame. Combines aspects of NeuroMaps Atlas | Online for download |
| 112RM-SL [41] | Rhesus macaque brain | T1 and T2 anatomical templates | Online for download |
| Visible Mouse Atlas [42] | Mouse (MR) | Effective volume resolution down to 100 × 100 × 100 μm ($1 \times 10^{-3}$ mm$^3$) | Umlaut Software |
| Scalable Brain Atlas [43] | Various | Fully web-based display engine for brain atlases | Online viewing |
| Duke/Calabrese MRI + DTI Rhesus Macaque Atlas [44] | Rhesus macaque brain | Diffusion tensor MRI atlas based on postmortem scans of ten rhesus macaques. Detailed segmentation with 242 anatomical structures | Online for download |
| MRI/DTI-Based Ferret Brain Atlas [45] | Ferret brain | Population-based MRI and DTI templates of adult ferret brain and tools for voxel-wise analysis; in vivo and ex vivo; 48 regions | Freely available |
| Valdés-Hernández MR Rat Brain Templates [46] | Rat brain | T2 MRI template set for morphometry, tissue segmentation, and fMRI localization in rats; includes white and gray matter probabilistic segmentations; fMRI | Freely available |
| Waxholm Space Atlas [47] | Sprague Dawley rat brain | MRI/DTI volumes in NIfTI ures | Freely available |

**Table 1.2** (continued)

| Atlas | Species | Construction (voxel/surface, automatic/manual) | Availability |
|---|---|---|---|
| T1 MRI Sheep Atlas [48] | Sheep | Population-averaged T1 MR template—gray matter, white matter, and CSF | Freely available |
| Digital Atlas of Dog Brain [49] | Mixed-breed dog | 1 mm and 0.33 mm isotropic diffeomorphic templates of the canine brain, as well as cortical surface representations suitable for use in FreeSurfer. In vivo, ex vivo, and surface templates. T1 MR | Free to use, but cannot be distributed for commercial gain |

can be particularly noticeable in brain applications that require separation of gray matter (GM), white matter (WM), and cerebrospinal fluid (CSF). In this situation, the application of multimodal data and the previously described multi-atlas segmentation method provide powerful tools for improving quantitation. By imaging representative subjects with both CT and anatomical MRI within a short temporal window (such that anatomy is constant within subject) and segmenting the GM, WM, and CSF from the MRI data sets, a reference library may be constructed for these regions. When a new PET/CT or SPECT/CT data set is acquired, the MAS tool may be used with a reference library of CT data sets but MR-derived atlas regions to build subject-specific segmentations of GM, WM, and CSF even in the absence of soft tissue contrast.

Consider an example study in which GM, WM, and CSF PET signals were to be evaluated longitudinally in several cohorts of mice acquired using a dedicated PET/CT scanner. Prior to execution of the PET study, the MR and CT reference library data were constructed. High-resolution T2-weighted MR images (Biospec 7 T/20 cm, Bruker, T2-weighted RARE, TR = 2600 ms, TE = 74 ms, 10 averages, 15 1-mm slices, FOV = 1.7 × 1.6 cm, matrix = 128 × 128), with co-registered CT images (Inveon PET/CT, Siemens, 80 kVp, 500 uA, 215 ms per projection, 0.21 × 0.21 × 0.21 mm voxel size using filtered back projection with Shepp filter at Nyquist point), were acquired in 12 healthy wild-type mice. GM, WM, and CSF regions were manually segmented. Example segmentations from three coronal slices, displayed as probability maps in the template space, are shown in Fig. 1.4a. Using a leave-one-out approach, the GM/WM/CSF regions were segmented for each of the individual MR and CT data sets using the MAS methodology. The left panel of Fig. 1.4b shows a coronal MR slice for the individual mouse. The middle panel shows the probability map for the WM region, generated using MR data (the gold standard). The right panel shows the same probability map generated using CT data, which is largely devoid of soft tissue contrast. As shown in Fig. 1.4c, while segmentation performance was superior using MR data, Dice coefficients of nearly 0.8 were achieved using only CT data. This performance enabled successful application of a CT + ROI reference library to study PET/CT data for segmentation of complex

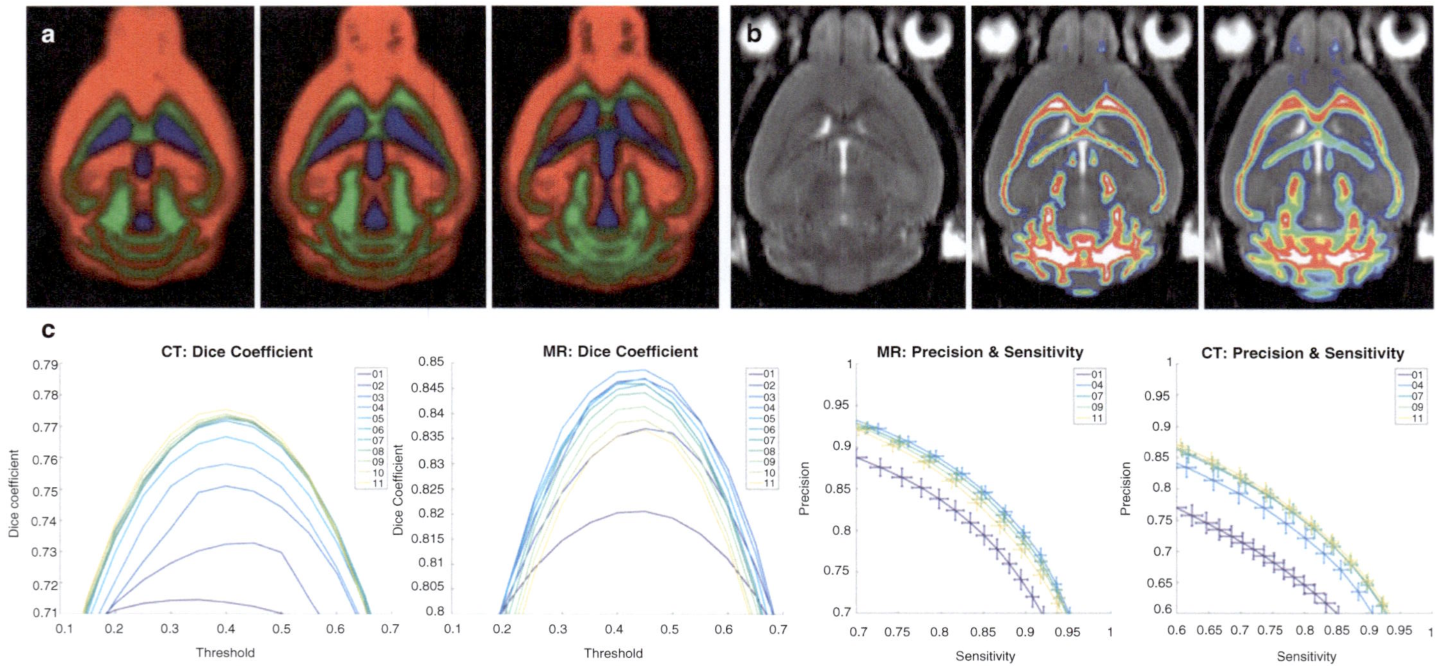


**Fig. 1.4** Segmentation of gray matter (GM), white matter (WM), and cerebrospinal fluid (CSF) regions from CT data is challenging due to the lack of soft tissue contrast. A cohort of 12 mice was imaged using MRI and CT, followed by expert manual segmentation of GM, WM, and CSF structures from the MRI. (**a**) Visualization of probability maps for GM (red), WM (green), and CSF (blue) regions from the cohort of 12 animals, following nonlinear co-registration to a template space. (**b**) The original MR (left) is segmented using a multi-atlas segmentation (MAS) approach to generate probability maps of the WM using (**b**) MRI data, and (**c**) CT data as the reference library. Even without the soft tissue contrast provided by MRI, a CT-only approach provides nearly as accurate segmentation of the WM region. (**c**) Comparisons of registration performance between MRI- and CT-based reference libraries in terms of Dice coefficient, sensitivity, and precision

brain structures in the absence of soft tissue contrast. Subsequently, this approach now has the utility of more confidently interrogating noninvasive methods targeting WM and/or CSF in addition to GM regions.

## 1.4 Combining Multimodal Data to Provide a Comprehensive Picture

In addition to their utility for extracting accurate information from images, multimodal data are pivotal in their ability to inform study outcomes by spanning multiple orders of magnitude both spatially and temporally. Software tools are critical in this capacity for visualization and image processing.

### 1.4.1 Image Fusion Visualization

MRI is the gold standard for 3D segmentation; however, even CT can be particularly helpful for segmentation of soft tissue organs, including the heart, lungs, and kidneys. In preclinical research, this is particularly important when one treatment group shows very low levels of PET or SPECT uptake in a region of interest. Such an organism can be invisible on PET/SPECT but still identifiable in CT/MRI.

The fusion process is used to visualize an overlay between two or more images. It is one way to qualitatively assess the co-registration; however, it is also a useful technique for a variety of research applications. In tandem, multimodal images can be used to simultaneously outline different features of an image. For example, a PET image can be used to clearly identify a xenograft tumor, while its co-registered CT image delineates other anatomical features, such as the lungs, heart, and kidneys.

A number of implementation details play an important role in how two- and three-dimensional intensity matrices are converted to red, green, and blue pixel values on a display. Most reconstructed image formats define a physical patient coordinate system in which distances can be measured in millimeters. For most simultaneous multimodal scanners, this coordinate system is shared across modalities and clearly defined in the reconstructed image headers, allowing the software to co-register the images without any transform optimization. In the event that images without aligned coordinate systems are loaded, the software should be able to reasonably display them together, for example, by reasonably translating one image coordinate system such that it is centered in the other image's field of view. Even when image coordinate systems are physically aligned, multimodal pairs of images will rarely share the same number of slices, voxel sizes, or field of view. For fused cross-sectional slice views, slices from each image are sampled along the same two-dimensional plane in physical coordinates, and each resulting sample is fused across images. In some preclinical MRI applications, the resolution along one axis will be significantly different to the other two. When these thick slices are sampled, and fused, any cross section along the low-resolution axis will appear much coarser.

Resampling an image with anisotropic voxels is necessary for visualization but may be undesirable for analysis. Image Registration Pitfalls describes how resampling may introduce quantitation errors. In addition, resampling an image to be isotropic in memory may greatly increase the total voxel count and correspondingly increase the runtime required for most analysis routines without increasing the quality or information content in the output. Where possible, anisotropic images should be filtered with physical distances and neighborhoods or processed with 2D routines on each high-resolution slice.

The typical color transfer function in medical imaging has a simple linear form. That is, a defined array of colors is indexed linearly based on the voxel intensity. It should be noted that equivalent differences along the color array indices correspond approximately to equivalent differences in voxel intensity, within error due to color discretization. In research applications, parametric derived images or autoradiography images are examples where a logarithmic scale may be more appropriate for its color transfer function. Individual image visualization and especially image fusion are most informative when intensity contrasts of interest map to a linear difference in color. If a linear color scale is used for logarithmically scaled intensity values, the image will appear to have little to no contrast and may saturate the image relative to others when it is fused.

Once intensity values are resampled and mapped to colors, image fusion is executed with a chosen blending algorithm. The most common blending technique is alpha blending, a linear combination between each color channel parameterized by their corresponding opacity values. The higher the alpha value, the larger that image's contribution will be to the final blended pixel color. When fusing more than two images together, a typical implementation will fuse the first two, fuse the result with the third, and so on. The alpha compositing process implies that the final color can depend on the order in which the three colors are blended. DWI is a special case where it is common for the three components of the voxel's tensor vector to be mapped linearly to red, green, and blue channels of a final fused image.

#### 1.4.1.1 Volume Rendering

Cross-sectional views provide 2D slices of 3D image matrices and form the standard views for clinical assessments. In research and preclinical imaging, it is also customary for software to provide 3D graphical projections to give visualizations a 3D appearance on the screen. In a traditional 3D graphics scene, objects are represented with surface-based objects with particular reflectance and lighting properties. Ray casting is used to project the appearance of these surfaces to a 2D representation on the screen [51]. Ray casting is also used for volume rendering, albeit with a slightly different approach. Instead of intersecting with surfaces on the screen, rays are projected into the image volume, and a resulting projected pixel color is a function of all voxels in the volume intersected by the ray. A maximum intensity projection computes the maximum voxel intensity along the ray and maps this value through the image's color transfer function to produce the visible color. This has the effect of rendering the "brightest" features within the volume. Similarly, color mappings along the said ray can be composited from pixel to pixel during traversal with a compositing function that can be optimized to highlight particular

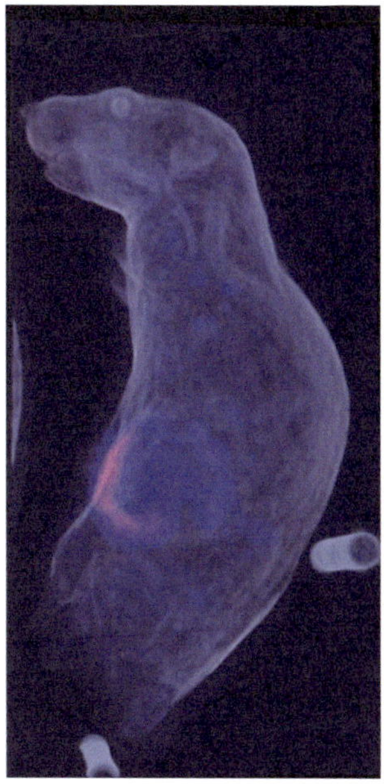

**Fig. 1.5** Fused volume renderings of 3D reconstructed fluorescence images and white light images using composite volume rendering

tissue intensities. An example of the fusion between two composite renderings is shown in Fig. 1.5.

Whereas cross-sectional views only allow traversal along the plane's perpendicular axis, a projection can be manipulated in all three dimensions by transforming the 3D volume in the scene prior to ray casting. The ray casting algorithm itself is highly parallelizable, allowing its implementation to be executed on modern graphics hardware with interactive performance. Graphic processing units (GPU) typically have their own dedicated memory hardware, often with less capacity than random access memory on the machine. This makes GPU-accelerated rendering of particularly large image volumes intractable on most computers without prior down-sampling done by the software. The Visualization Toolkit provides rendering implementations both on the CPU and the GPU for maximal cross-platform compatibility.

Fusion for volume rendering is typically done using alpha blending of the 2D projections from multiple images. Algorithmically computed opacity values can often enhance the appearance of surface boundaries within the volume when the gradient is used.

## 1.5 Applications of Preclinical Image Fusion

One of the primary advantages of multimodal imaging is the generation of complementary imaging around a common biological question. The ability to combine data from various sources of imaging information is an area of active research and can depend strongly on the specific task or question. In this section, we describe examples from three scenarios. First, multimodal data across scales can be compared directly. Second, multimodal data across disparate scales or types can be combined to tell a more complete biological story. Third, data from multiple modalities can be incorporated into complex models, often using machine learning methods to improve our ability to perform classification and estimation tasks.

### 1.5.1 Cryofluoresence Tomography (CFT)

For example, consider the case of fluorescence imaging. A fluorophore may be administered intravenously or intrathecally and imaged longitudinally in vivo to provide some low-resolution spatial distribution information as well as in vivo temporal information about biodistribution. That same subject can then be sacrificed, and organs can be prepared for subsequent whole-body or focused organ ex vivo CFT to provide higher-resolution distribution information at a macroscopic level. During CFT acquisition, individual slices may be transferred to tape or slide to enable fluorescence microscopy. Depending on the compounds and studies of interest, this approach can be extended to other modalities. For example, if a compound can be labeled with both a PET tracer and a fluorophore, then high temporal resolution PET dynamic imaging can be used to support pharmacokinetic modeling applications through in vivo imaging prior to higher-resolution, single timepoint ex vivo CFT imaging. Alternately, for even higher-resolution imaging of focused regions, transfer of individual sections to the appropriate medium enables the use of conventional immunohistochemistry (IHC) staining. Consider the intrathecal administration of a Cy7-conjugated compound. A subject's brain is blocked and imaged via CFT to produce a high-resolution brain-wide distribution of signal as shown in Fig. 1.6. Individual sections are transferred to tape for IHC staining as shown in Fig. 1.6. Using these approaches, a single subject may be used to generate temporal, macroscopic, and microscopic data—such approaches maximize scarce resources while minimizing variability.

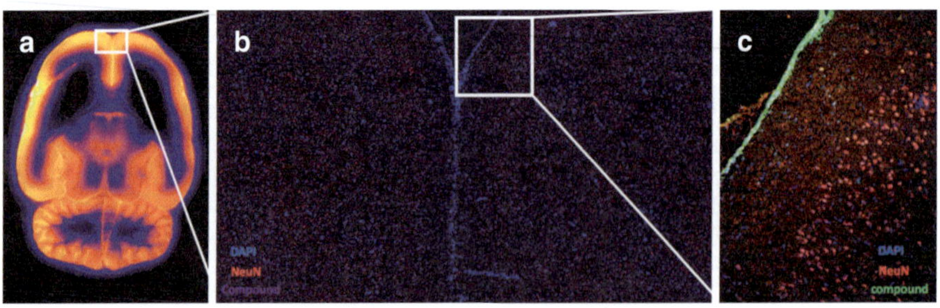


**Fig. 1.6** The ability to utilize a single animal for imaging across multiple scales is often beneficial as it removes intersubject variability. (**a**) Whole-brain CFT is used to visualize and assess fluorophore distribution at the whole-brain level. Transfer of individual sections during CFT processing enables application of fluorescence microscopy in specific subsections of the brain as shown here at (**b**) ~30× magnification, and (**c**) ~40× magnification

As another example, consider a surgery study in hounds for study of a pancarpal arthrodesis model. In this study, two cohorts of five hounds received different concentrations of a bone morphogenetic protein (BMP) co-labeled to $^{124}$I and $^{125}$I. In vivo $^{124}$I PET/CT imaging of the paw region was performed out to 2 weeks post-surgery. $^{125}$I planar imaging was performed in vivo out to 5 weeks. Through multiple image processing and calibration steps, the percent injected activity in multiple paw regions could be combined across the entire 5-week in vivo imaging period to enable more accurate estimates of residence times and other cohort comparisons. After 5 weeks, ex vivo 3D quantitative autoradiography and accompanying white light imaging was performed. This imaging approach enabled quantitative analysis and qualitative visualization of BMP distribution at high resolution as shown in Fig. 1.7. The combination of these far-reaching multimodal imaging techniques yielded a richer data set and greater statistical power from ten canines when compared to a previous non-imaging study that utilized fifty canines.

Utility can also be gained from direct quantitative comparison of multimodal imaging data across different spatial scales. For example, in vivo SPECT and PET imaging are commonly used to estimate drug concentration within the brain.

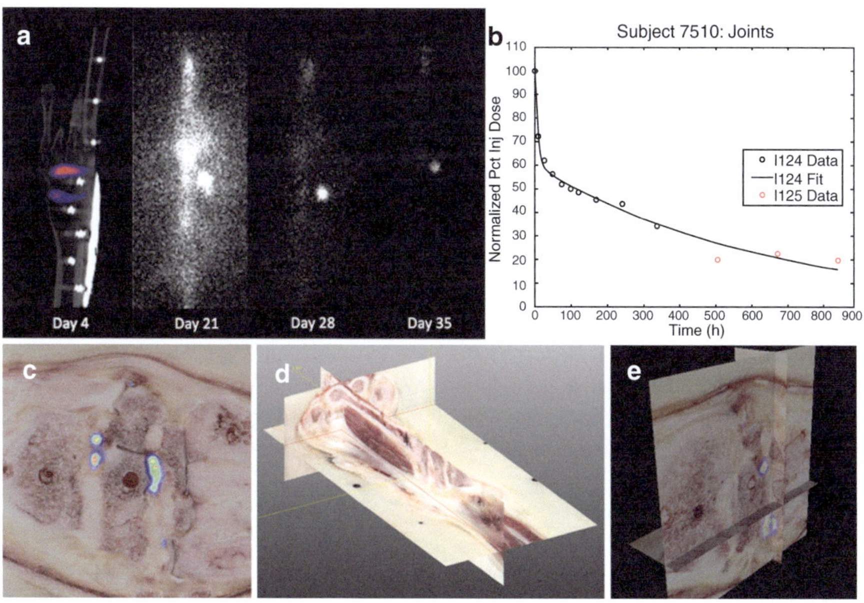


**Fig. 1.7** A combined in vivo/ex vivo imaging pipeline provides quantitative and qualitative information to describe the behavior of a BMP in a canine pancarpal arthrodesis model. (**a**) In vivo PET/CT (PET in purple/blue, CT in gray) and planar imaging are utilized to estimate $^{124}$I and $^{125}$I signal, respectively, out to 5 weeks post-surgery. (**b**) Information from these two modalities is combined into a single quantitative time-activity curve, enabling pharmacokinetic analysis between cohorts. (**c**–**e**) 3D cryo-imaging quantitative autoradiography is combined with white light imaging to provide high-resolution qualitative localization of $^{125}$I signal post-sacrifice

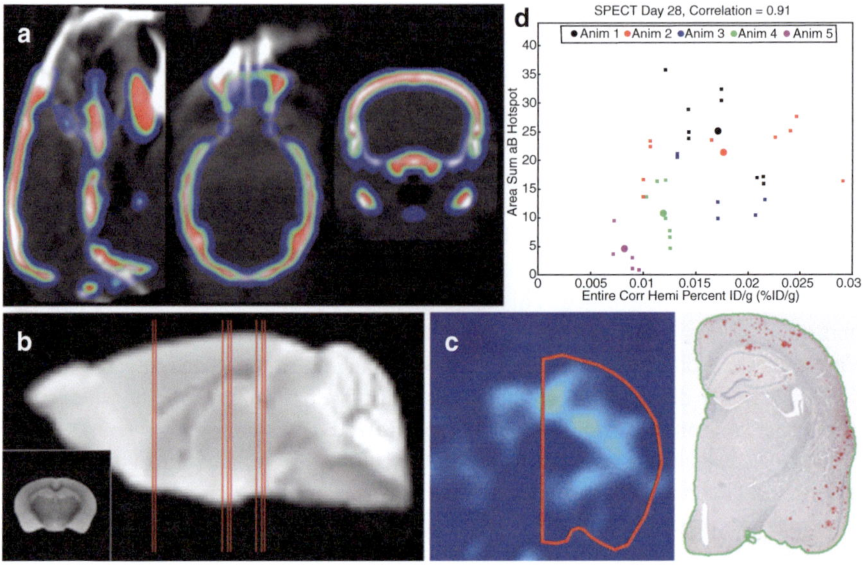


**Fig. 1.8** Pipeline to enable correlative analysis of quantitative SPECT and IHC data in the mouse brain. (**a**) Upon acquisition of SPECT/CT data, the test CT data are co-registered to a template CT. (**b**) A template MR exists in the space of the template CT. The Bregma coordinate system is utilized to identify the MR slice corresponding to each IHC section (vertical red lines). (**c**) The corresponding SPECT slice (left) is extracted, and its mean signal in %ID/g is compared to the corresponding IHC section's (right) total beta-amyloid burden. (**d**) SPECT and IHC summary statistics are compared across multiple animals and sections, providing a strong validation between estimation methods

Analysis of these distributions through ex vivo staining can provide useful validation but is often limited to visual, qualitative comparisons. In [52], in vivo SPECT/CT images were co-registered and aligned to a template space. In the template space, individual slices at different stereotaxic coordinates were extracted, and a total %ID/g value within a hemisphere was computed. This %ID/g value was directly compared to the estimates of plaque load derived via IHC. Hemispheric estimates of IHC Aβ load and SPECT %ID/g were linearly related with an $R^2 = 0.83$ as shown in Fig. 1.8.

## 1.5.2 Integration of Multimodal Imaging with Machine Learning

Computational tools derived from machine learning research have had great success in image segmentation, data classification, and multimodal processing applications [53, 54]. Machine learning is increasingly featured in imaging journals and conferences with applications ranging from diagnosis to segmentation and longitudinal prediction [55–57]. Imaging itself offers a variety of modalities which can be further integrated with data from genomics and other biological assays to create sophisticated models of disease and biomarkers [58, 59]. Indeed, the goal of this active area of research is to leverage multimodal data at a larger scale to derive new insights moving forward.

Machine learning models can be trained in a supervised manner when ground truth results are available or in an unsupervised manner for blind discovery of underlying patterns in the data. Image segmentation is an example where supervised models have been trained with great success. Large software companies have contributed methods and software tools which they use at large scales internally for image classification and processing [60, 61]. Challenges have arisen in the application of these methods to medical images [57]. 2D convolutional neural networks do not leverage spatial correlations in the third dimension of typical 3D matrix sizes. Furthermore, when 3D convolution kernels are used, the computational resource demands exceed the capacity of individual workstations and do not scale efficiently with increasing kernel sizes. Nevertheless, a number of applications have been discovered where these models outperform classical methods [62]. The disadvantage of some of these "deep learning" approaches can be that the highly parameterized model does not take full advantage of relevant domain knowledge and may not shed much light on the image generation process itself.

## 1.6 Conclusion

Preclinical multimodal data have tremendous utility. The use of the appropriate software tools and resources, such as atlases, is critical to practical use of multimodal images for the extraction of quantitative data and for the combination of data to facilitate meaningful interpretation in the context of complex biological questions. Available software tools are well-designed for handling of multimodal data across multiple temporal and spatial scales. These tools have been specifically designed to enable seamless visualization as well as to support an array of image processing operations. Currently, the most important of these operations as it pertains to multimodal data is image registration with mature methods supporting a variety of tasks and data types. As illustrated in the examples throughout this chapter, such tools are already aiding in resource optimization, ethical use of animals through a reduction in necessary animal number, and improved data utilization and outcome measures. The burgeoning fields of machine learning and artificial intelligence are rapidly growing in importance for multimodal analysis and will further increase the utility of preclinical multimodal data as they transition from academia-heavy development environments to more turn-key commercial availability. As imaging continues to cement itself within the drug development cycle from discovery through late phase clinical trials, the need for and uses of multimodal data will continue to expand. In support of these needs, it is critical that innovation around software tools for image fusion continues and performs robustly across different species and disease models.

## References


1. Kinahan PE, Townsend DW, Beyer T, Sashin D. Attenuation correction for a combined 3D PET/CT scanner. Med Phys. 1998;25:2046–53.
2. Fox GB, Chin C-L, Luo F, et al. Translational neuroimaging of the CNS: novel pathways to drug development. Mol Interv. 2009;9:302.
3. Invicro VivoQuant. http://www.vivoquant.com. Accessed 2 Nov 2017.
4. Schindelin J, Arganda-Carreras I, Frise E, et al. Fiji: an open-source platform for biological-image analysis. Nat Methods. 2012;9:676–82. https://doi.org/10.1038/nmeth.2019.
5. SPM software—Statistical Parametric Mapping. http://www.fil.ion.ucl.ac.uk/spm/software/. Accessed 2 Nov 2017.
6. Jenkinson M, Beckmann CF, Behrens TEJ, et al. FSL. NeuroImage. 2012;62:782–90. https://doi.org/10.1016/j.neuroimage.2011.09.015.
7. Loening AM, Gambhir SS. AMIDE: a free software tool for multimodality medical image analysis. Mol Imaging. 2003;2:131–7.
8. AnalyzeDirect.com. https://analyzedirect.com/. Accessed 2 Nov 2017.
9. Cox RW. AFNI: software for analysis and visualization of functional magnetic resonance neuroimages. Comput Biomed Res Int J. 1996;29:162–73.
10. Ayachit U. The ParaView guide: updated for ParaView version 4.3, Full color version. Los Alamos: Kitware; 2015.
11. Yushkevich PA, Piven J, Cody Hazlett H, et al. User-guided 3D active contour segmentation of anatomical structures: significantly improved efficiency and reliability. NeuroImage. 2006;31:1116–28.
12. PMOD Technologies LLC—PMOD Technologies. https://www.pmod.com. Accessed 2 Nov 2017.
13. Kikinis R, Pieper SD, Vosburgh KG. 3D slicer: a platform for subject-specific image analysis, visualization, and clinical support. In: Jolesz FA, editor. Intraoperative imaging and image-guided therapy. New York, NY: Springer; 2014. p. 277–89.
14. FreeSurfer. http://surfer.nmr.mgh.harvard.edu/. Accessed 3 Nov 2017.
15. Amira 3D Software for Life Sciences|Thermo Fisher Scientific. https://www.fei.com/software/amira-for-life-sciences/. Accessed 3 Nov 2017.
16. Rosset A, Spadola L, Ratib O. OsiriX: an open-source software for navigating in multidimensional DICOM images. J Digit Imaging. 2004;17:205–16.
17. Avants BB, Tustison NJ, Stauffer M, et al. The insight ToolKit image registration framework. Front Neuroinform. 2014;8:44. https://doi.org/10.3389/fninf.2014.00044.
18. VTK—The Visualization Toolkit. https://www.vtk.org/. Accessed 2 Nov 2017.
19. Open Inventor®|Open Inventor 3D SDK. http://www.openinventor.com/. Accessed 3 Nov 2017.
20. Qt|Cross-platform software development for embedded & desktop. https://www.qt.io/. Accessed 3 Nov 2017.
21. Java Software|Oracle. https://www.oracle.com/java/index.html. Accessed 3 Nov 2017.
22. Dubois A, Hérard A-S, Delatour B, et al. Detection by voxel-wise statistical analysis of significant changes in regional cerebral glucose uptake in an APP/PS1 transgenic mouse model of Alzheimer's disease. NeuroImage. 2010;51:586–98.
23. Lebenberg J, Hérard A-S, Dubois A, et al. Validation of MRI-based 3D digital atlas registration with histological and autoradiographic volumes: an anatomofunctional transgenic mouse brain imaging study. NeuroImage. 2010;51:1037–46.
24. Worsley KJ, Liao CH, Aston J, et al. A general statistical analysis for fMRI data. NeuroImage. 2002;15:1–15.
25. Murgasova MK, Lockwood-Estrin G, Nunes RG, et al. Distortion correction in fetal EPI using non-rigid registration with a Laplacian constraint. IEEE Trans Med Imaging. 2018;37(1):12–9.
26. Hédouin R, Commowick O, Bannier E, et al. Block-matching distortion correction of echo-planar images with opposite phase encoding directions. IEEE Trans Med Imaging. 2017;36:1106–15.

27. Jezzard P, Clare S. Sources of distortion in functional MRI data. Hum Brain Mapp. 1999;8:80–5.
28. Klein A, Andersson J, Ardekani BA, et al. Evaluation of 14 nonlinear deformation algorithms applied to human brain MRI registration. NeuroImage. 2009;46:786–802.
29. Cheng J, Qiu W, Yuan J, et al. Accurate quantification of local changes for carotid arteries in 3D ultrasound images using convex optimization-based deformable registration. In: Medical imaging: image processing. Proc SPIE. 2016;9784:978448.
30. Elfarnawany M, Alam SR, Agrawal SK, Ladak HM. Evaluation of non-rigid registration parameters for atlas-based segmentation of CT Images of Human Cochlea. In: SPIE medical imaging. International Society for Optics and Photonics. Proc SPIE. 2017;10133:101330Z.
31. Pierce LA, Elston BF, Clunie DA, et al. A digital reference object to analyze calculation accuracy of PET standardized uptake value. Radiology. 2015;277:538–45.
32. Erlandsson K, Buvat I, Pretorius PH, et al. A review of partial volume correction techniques for emission tomography and their applications in neurology, cardiology and oncology. Phys Med Biol. 2012;57:R119.
33. Dogdas B, Stout D, Chatziioannou AF, Leahy RM. Digimouse: a 3D whole body mouse atlas from CT and cryosection data. Phys Med Biol. 2007;52:577–87. https://doi.org/10.1088/0031-9155/52/3/003.
34. Khmelinskii A, Groen HC, Baiker M, et al. Segmentation and visual analysis of whole-body mouse skeleton microSPECT. PLoS One. 2012;7:e48976.
35. Nie B, Chen K, Zhao S, et al. A rat brain MRI template with digital stereotaxic atlas of fine anatomical delineations in paxinos space and its automated application in voxel-wise analysis. Hum Brain Mapp. 2013;34:1306–18.
36. Lein ES, Hawrylycz MJ, Ao N, et al. Genome-wide atlas of gene expression in the adult mouse brain. Nature. 2007;445:168–76.
37. Segars WP, Sturgeon G, Mendonca S, et al. 4D XCAT phantom for multimodality imaging research. Med Phys. 2010;37:4902–15.
38. Paxinos G, Franklin KB. The mouse brain in stereotaxic coordinates. Houston, TX: Gulf Professional Publishing; 2004.
39. Rohlfing T, Kroenke CD, Sullivan EV, et al. The INIA19 template and NeuroMaps atlas for primate brain image parcellation and spatial normalization. Front Neuroinform. 2012;6:27.
40. Frey S, Pandya DN, Chakravarty MM, et al. An MRI based average macaque monkey stereotaxic atlas and space (MNI monkey space). NeuroImage. 2011;55:1435–42.
41. McLaren DG, Kosmatka KJ, Oakes TR, et al. A population-average MRI-based atlas collection of the rhesus macaque. NeuroImage. 2009;45:52–9.
42. Petiet AE, Kaufman MH, Goddeeris MM, et al. High-resolution magnetic resonance histology of the embryonic and neonatal mouse: a 4D atlas and morphologic database. Proc Natl Acad Sci. 2008;105:12331–6.
43. Bakker R, Tiesinga P, Kötter R. The scalable brain atlas: instant web-based access to public brain atlases and related content. Neuroinformatics. 2015;13:353–66.
44. Calabrese E, Badea A, Coe CL, et al. A diffusion tensor MRI atlas of the postmortem rhesus macaque brain. NeuroImage. 2015;117:408–16.
45. Hutchinson EB, Schwerin SC, Radomski KL, et al. Population based MRI and DTI templates of the adult ferret brain and tools for voxelwise analysis. NeuroImage. 2017;152:575–89.
46. Valdes Hernandez PA, Sumiyoshi A, Nonaka H, et al. An in vivo MRI template set for morphometry, tissue segmentation, and fMRI localization in rats. Front Neuroinform. 2011;5:26.
47. Papp EA, Leergaard TB, Calabrese E, et al. Waxholm space atlas of the sprague dawley rat brain. NeuroImage. 2014;97:374–86.
48. Nitzsche B, Frey S, Collins LD, et al. A stereotaxic, population-averaged T1w ovine brain atlas including cerebral morphology and tissue volumes. Front Neuroanat. 2015;9:69.
49. Datta R, Lee J, Duda J, et al. A digital atlas of the dog brain. PLoS One. 2012;7:e52140.
50. Wang H, Suh JW, Das SR, et al. Multi-atlas segmentation with joint label fusion. IEEE Trans Pattern Anal Mach Intell. 2013;35:611–23.
51. Glassner AS. An introduction to ray tracing. Palo Alto: Elsevier; 1989.

52. Patel S. SPECT imaging of an 125I-labeled antibody against amyloid plaques and determination of correlation with amyloid immunostaining. Poster presented at the 10th International symposium on functional neuroreceptor mapping of the living brain, May 2014, Egmond aan Zee, The Netherlands.
53. LeCun Y, Bengio Y, Hinton G. Deep learning. Nature. 2015;521:436–44.
54. Libbrecht MW, Noble WS. Machine learning applications in genetics and genomics. Nat Rev Genet. 2015;16:321–32.
55. Huynh T, Gao Y, Kang J, et al. Estimating CT image from MRI data using structured random forest and auto-context model. IEEE Trans Med Imaging. 2016;35:174–83.
56. Parmar C, Grossmann P, Bussink J, et al. Machine learning methods for quantitative radiomic biomarkers. Sci Rep. 2015;5:13087.
57. Chen J, Yang L, Zhang Y, et al. Combining fully convolutional and recurrent neural networks for 3D biomedical image segmentation. In: Lee DD, Sugiyama M, Luxburg UV, et al., editors. Advances in neural information processing systems 29. Red Hook, NY: Curran Associates, Inc.; 2016. p. 3036–44.
58. Batmanghelich NK, Dalca A, Quon G, et al. Probabilistic modeling of imaging, genetics and diagnosis. IEEE Trans Med Imaging. 2016;35:1765–79.
59. Yao X, Yan J, Kim S, et al. Genetic findings using ADNI multimodal quantitative phenotypes: a 2014 update. Alzheimers Dement J Alzheimers Assoc. 2015;11:P426.
60. Shotton J, Sharp T, Kipman A, et al. Real-time human pose recognition in parts from single depth images. Commun ACM. 2013;56:116–24.
61. Abadi M, Agarwal A, Barham P, et al (2016) Tensorflow: large-scale machine learning on heterogeneous distributed systems. arXiv:1603.04467
62. Prasoon A, Petersen K, Igel C, et al. Deep feature learning for knee cartilage segmentation using a triplanar convolutional neural network. In: Mori K, Sakuma I, Sato Y, Barillot C, Navab N, editors. Medical image computing and computer-assisted intervention—MICCAI 2013: 16th International Conference, Nagoya, Japan, September 22-26, 2013, Proceedings, Part II. Berlin: Springer; 2013. p. 246–53.

# Instrumentation Challenges in (S)PE(C)T Systems

2

David Brasse and Frederic Boisson

## 2.1 Introduction

To date, molecular imaging is globally recognized as a powerful tool to assess in vivo anatomical and functional structures within a living subject [1, 2]. A wide range of imaging modalities based on different physical principles is available. Used individually, the value of modalities such as X-ray computed tomography (CT), single-photon emission computed tomography (SPECT), positron emission tomography (PET), magnetic resonance imaging (MRI), and ultrasound (US) is well recognized, both in clinical and preclinical fields [3–12]. Although established for decades, these modalities are still subject to much research, both on the development of new detection components and the design of new probes to artificially improve the natural contrast induced by physical processes.

With different performances in terms of spatial and temporal resolutions, sensitivity, and specificity but also their ability to provide quantitative results, all of the previously mentioned imaging modalities allow to carry out a great variety of studies. However, the need for and relevance of obtaining a maximum of information relating to the same subject is undeniable. The last two decades have thus seen the advent of a full-fledged research portfolio on multimodal in vivo imaging systems, leading to specific instrument developments.

Despite the certain enthusiasm for multimodal imaging, it is important to highlight the instrumental developments specific to each imaging modality, in particular those concerning emission tomography modalities. Since the introduction of the Anger camera in the 1950s [13], instrumentation (hardware) has become one

D. Brasse (✉) · F. Boisson
Université de Strasbourg, CNRS, IPHC, Strasbourg, France
e-mail: david.brasse@iphc.cnrs.fr



C. Kuntner-Hannes, Y. Haemisch (eds.), *Image Fusion in Preclinical Applications*, https://doi.org/10.1007/978-3-030-02973-9_2

essential element for the design of imaging systems. The gamma photon detection principle based on the use of scintillating crystals, photomultiplier tubes (PMTs), and readout electronics has been the basis of almost all clinical and preclinical PET and SPECT systems. At the moment, we witness a rapid technology change in the area of photodetectors with more and more systems using semiconductor photodetectors such as avalanche photodiodes (APDs) [14, 15] and silicon PMs (SiPMs). The change has first been triggered by the attempt to combine PET and MRI, since PMTs are not compatible with magnetic fields. A second motivation for the transition to SiPM photodetectors in particular is their better timing behavior, important for the development of time-of-flight (TOF) (clinical) PET systems. For preclinical applications, their compactness and the ability to read out very small pixels make them particularly attractive. As the use of SiPMs increases, their costs are coming down to competitive levels. In the same way, semiconductor detectors based on CdTe or CdZnTe have emerged as the alternative of choice for SPECT system developments, thanks to their high energy and intrinsic spatial resolutions.

This chapter focuses on instrumentation developments for nuclear emission tomography imaging, without claiming completeness. It presents recent developments of PET and SPECT imaging technology by focusing on their origin triggered by scientific questions but also characteristics that are specific to each of these modalities. Thus, common notions to PET and SPECT imaging, such as the positioning of the interaction at the level of the detection module or the key elements of the acquisition chain, are discussed. Special attention will be paid to instrumentation developments aiming at an insertion of PET or SPECT systems into MRI systems.

## 2.2 Instrumentation in PET

At the early stage of this century, clinically available medical imaging technologies have been adapted for use in preclinical investigations. The tremendous effort in biological research focusing on the observation of the living at the molecular scale pushed forward the development of such dedicated positron emission tomography (PET) systems.

Techniques and methods used in molecular imaging combined with drug development gave rise to a wide range of biological processes observed and measured quantitatively using PET.

A number of important issues have been addressed while designing dedicated preclinical PET systems. The main challenges have been to optimize both detection efficiency and spatial resolution to correctly extract biological information from a single living animal model facilitating the bridge between basic science and clinical research.

This section presents an overview of instrumentation developments in PET in terms of detection module and readout strategies.

### 2.2.1 Basic Components of a Preclinical PET Module

PET imaging is based on the simultaneous detection of two opposite gamma rays generated by the annihilation of a positron with an electron within the volume of interest. The detected photons contribute to identify a coincidence event occurring in a volume described by the line joining the two triggered detectors called a line of response (LOR). Over the last 20 years, hardware and software methods have been proposed to improve the definition of each LOR (spatial resolution) and to increase both the number of LOR and the number of coincidences acquired inside each LOR (detection efficiency). PET molecular imaging dedicated to small-animal studies requires both high spatial resolution and high sensitivity [16].

In most recent imaging systems, the geometry of the detector module is based on a block detector structure. The following picture (Fig. 2.1) is an illustration of a block structure introduced by M Casey and R Nutt [17] where a matrix of 64 crystals is read by 4 individual photomultiplier tubes (PMTs). The same concept has been transferred to a detection module dedicated to preclinical PET systems where individual PMTs have been replaced by position-sensitive PMTs (PS-PMTs) or multi-anode PMTs (MA-PMTs). In this configuration, crystals are identified using a flood histogram.

This histogram image is obtained experimentally by irradiating the surface of the detector with 511 keV photons. The location of each detected event is then calculated and stored in a two-dimensional image as illustrated in Fig. 2.2 for a matrix of 27 × 26 crystals read by MA-PMT.

The role of the scintillating crystal is to convert the high energy photons impinging the detector to a number of optical photons proportional to the deposited energy. Scintillating crystals can be organic or inorganic compounds and are characterized

**Fig. 2.1** ECAT EXACT HR+ (Siemens) detector block composed of 64 crystals and read by 4 PMTs

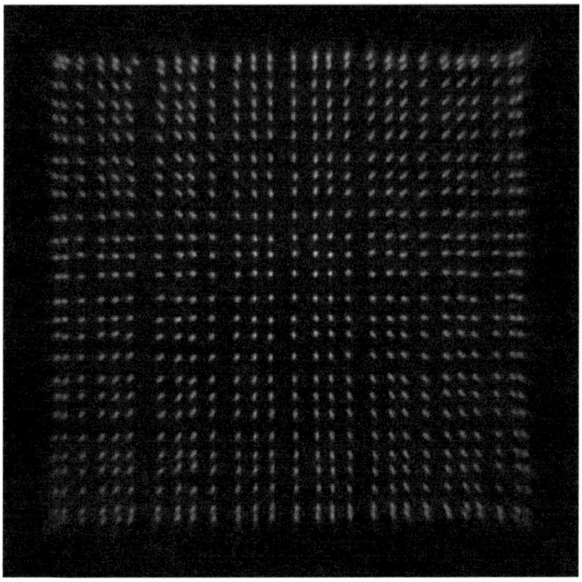

**Fig. 2.2** Illustration of a flood histogram obtained on an IRIS PET module composed of 27 × 26 LYSO crystals, 1.7 mm pitch, read by a H8500 MA-PMT (Image courtesy of Liji Cao, Inviscan)

by intrinsic properties such as their stopping power, their light yield, and their scintillation decay. The main requirements for crystals used in preclinical PET modules are (1) a high efficiency to convert the excitation energy to fluorescent radiation, (2) a transparency to its fluorescent radiation to optimize the transmission of light, (3) the emission of light in a spectral range detectable by photodetectors, (4) a short decay time to achieve fast response, and (5) a good radiation hardness.

Scintillating crystals used for preclinical PET are inorganic, mainly cerium-doped and lutetium-based. One drawback using lutetium-based crystals is the fact that 2.6% of naturally occurring lutetium is $^{176}Lu$, a long-lived radioactive isotope including beta and gamma decays resulting in a small natural radiation background increasing random events during PET acquisition. This radiation baseline can however be beneficial to, for example, energy calibration or daily quality control without any external radioactive source [18].

Table 2.1 lists the principal properties of bismuth germinate ($Bi_4Ge_3O_{12}$ or BGO), cerium-doped gadolinium oxyorthosilicate ($Gd_2SiO_5[Ce]$ or GSO), cerium-doped lutetium oxyorthosilicate ($Lu_2SiO_5[Ce]$ or LSO), cerium-doped lutetium orthoaluminate ($LuAlO_3[Ce]$ or LuAP), cerium-doped lutetium yttrium orthosilicate ($Lu_{2(1-x)}Y_{2x}SiO_5[Ce]$ or LYSO), and cerium-doped mixed lutetium and gadolinium orthosilicate ($Lu_{1.8}Gd_{0.2}SiO_5[Ce]$ or LGSO) crystals. Additional information on scintillators can be found in [19, 20].

Historically, PET scanners used photomultiplier tubes to convert the scintillation light into an electric current. The basic principle of a PMT is the multiplication of photoelectron with a series of dynodes in a vacuum environment. The scintillation light is transmitted through an entrance window of the vacuum tube and interacts with a thin layer of material called photocathode. Each optical

**Table 2.1** Properties of main scintillating crystals used for preclinical PET imaging

| | BGO | GSO | LSO | LuAP | LYSO | LGSO |
|---|---|---|---|---|---|---|
| Density (g $cm^{-3}$) | 7.13 | 6.70 | 7.40 | 8.34 | 7.11 | 7.3 |
| Effective atomic number | 75 | 59 | 66 | 65 | 65 | 63 |
| $\mu$ at 511 keV ($cm^{-1}$) | 0.89 | 0.72 | 0.88 | 0.95 | 0.82 | 0.80 |
| Photofraction (%) | 37.2 | 24.9 | 32.4 | 30.4 | 31.1 | 34 |
| Index of refraction | 2.15 | 1.89 | 1.81 | 1.94 | 1.81 | 1.8 |
| Light yield (ph $MeV^{-1}$) | 8200 | 9000 | 25,000 | 11,300 | 33,000 | 41,000 |
| Scintillation decay (ns) | 300 | 56 | 40 | 184 | 41 | 31–47 |
| $\lambda_{max}$ (nm) | 480 | 430 | 420 | 365 | 420 | 420 |

photon can then produce a photoelectron via photoelectric effect. The photoelectron is then accelerated by a high potential difference and focalized on a first dynode. The acquired energy during the acceleration process allows the next dynode, coated with an emissive material, to release three to four new electrons when the photoelectron interacts with it. After several stages of dynode amplification, each initial photoelectron gives birth to $10^6$ electrons. A PMT can be assumed as a current source where an optical photon can generate a charge of 160 pC in few nanoseconds.

An alternative to PMTs is the avalanche photodiodes (APDs). This photodetector is based on a silicon photodiode. The absorption of an incident optical photon generates electron-hole pairs. An applied reverse bias voltage on the junction creates a strong internal electric field which accelerates the electron through the diode and produces secondary electrons by impact ionization. The resulting electron avalanche can produce up to several hundreds of electrons.

More recently, SiPMs have been implemented as photodetectors for PET [21]. Further capitalizing on the idea of semiconductor avalanche photodiodes, SiPMs are matrices of miniaturized APDs (called "single-photon avalanche diodes" (SPADs)) operating in Geiger mode, showing the same compactness and non-sensitivity to magnetic field. Due to the Geiger-mode operation, SiPMs provide electronic gains up to $10^6$, independent of the incident signal and thus the ability to count single photons. The number of SPADs forming a SiPM determines its dynamic range, the size and density of SPADs, and the sensitivity and timing behavior of the SiPM. The linearity of a SiPM response is strongly correlated with the number of SPADs in a single pixel. Therefore, in a properly designed configuration, the number of SPADs corresponds to the expected number of optical photons produced by the scintillation process and transferred to the photodetector. In order to assess the energy, the remaining non-linearities have to be corrected [22, 23]. Optimizing both the energy and timing resolution requires to deal with the trade-off between the size and number of SPADs within a single SiPM pixel. It is interesting to observe the evolution of this emerging technology, particularly from the perspective of the crystal choice and their dimensions.

In this "analog" SiPM, the signals from individual SPADs are summed up, and the energy and timing information of each event are mainly derived from leading

**Table 2.2** Properties of main analog photodetectors used for preclinical PET imaging

| | PMT | APD | SiPM |
|---|---|---|---|
| Photodetection efficiency at 420 nm | 20–40% | 60–70% | 25–75% |
| Gain | $10^6$–$10^7$ | $10^1$–$10^3$ | $10^5$–$10^6$ |
| Rise time (ns) | ~1 | ~5 | ~1 |
| Bias (V) | >800 | 300–1000 | 30–80 |
| Temperature sensitivity (%/K) | <1 | ~3 | 1–8 |
| Compactness | – | +++ | +++ |
| Magnetic field sensitivity | – | +++ | +++ |

edge discrimination and signal integration. Digital approach has been introduced where every photoelectron detected in a SPAD is recorded with its own timing information. This digital process was first commercialized by Philips [24] where they only introduced one time-to-digital converter for a large number of SPAD. The number of triggered SPADs represents the energy of the event.

Table 2.2 summarizes the principle properties of the photodetectors used in preclinical PET systems to convert the optical photon flux to an electric signal.

The use of inorganic scintillating crystal coupled to photodetector is an indirect approach to convert 511 keV to electric signal. Another solution is to use a semiconductor material such as cadmium zinc telluride (CZT) to directly convert the absorbed energy of the gamma photon into a significant electronic signal. CZT is a wide band gap compound semiconductor with a relatively high effective atomic number. The electric signal resulting from the gamma interaction is read out by a network of metallic electrodes. Several parameters can be optimized to maximize the performance of such detector for PET imaging such as the width of the cathode strip, the steering strip, and the bias voltage [25]. Based on these developments, design of preclinical PET scanner has been proposed targeting high performance [26–29].

### 2.2.2 Event Positioning

In the block detector approach, the size of the individual crystal element and the light readout scheme determine the intrinsic spatial resolution [30]. The crystal segmentation is driven by several parameters to improve the spatial resolution of the detection module. It could be driven either by the photodetector segmentation in a one-to-one coupling scheme or by the limits of crystal identification when a light sharing approach is preferred.

A complete PET system consists in a large number of block detectors positioned around the animal to be imaged. The most common geometrical configuration of a preclinical PET system follows the ring geometry. The number of PET modules in the transverse plane defined the polygonal shape of the scanner, while the number of rings determines the axial extent of the imaging system.

The detection efficiency depends on the geometrical solid angle acceptance and the density and thickness of the detector material. To improve detection efficiency, thicker crystal elements can be used. However, the trade-off is a reduction in spatial resolution through parallax errors due to the lack of an accurate measurement of the depth of interaction (DOI).

To overcome this issue, a possible solution is to employ a so-called phoswich detector with two or more crystal layers sampling the interaction depth and thus obtaining an estimate of the DOI [31]. The DOI is discretized based on the number of layers used. The number of layers can be limited by the proper identification of each individual crystal. Several approaches have been investigated dealing with the offset of the layers in the flood map or with the intrinsic properties of the crystal composing each layer such as the scintillation decay or the concentration of the dopant [32, 33].

As an example, Ito et al. demonstrated the capability of identifying up to four layers [34].

For a continuous sampling of the DOI and a single-ended readout of the detector, DOI information can be extracted from the pulse shape. A dedicated coating is applied on the crystal surface resulting in a modification of the decay time of the detected pulse in a depth-dependent manner [35]. Several groups have used scintillator array coupled at both sides to avalanche photodiode (APD) arrays [36, 37], to position-sensitive APDs [38–41], or using hybrid solution [42–45]. In this configuration, the DOI resolution is mainly driven by the light shared on both photodetector sides.

Already in 1978, Ter-Pogossian et al. developed an original concept where thallium-doped sodium iodide (NaI(Tl)) crystals were positioned in the axial direction with a PMT at each end of the crystals [46]. The main advantage of this geometry is that neither resolution nor sensitivity is compromised due to a simplification in the depth encoding procedure. Figure 2.3 illustrates this geometrical approach.

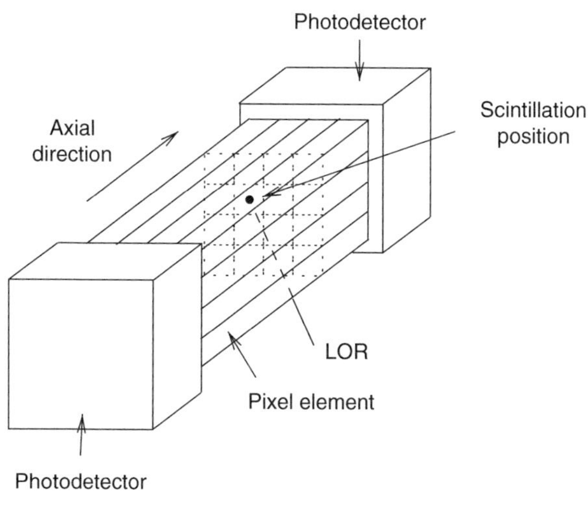


**Fig. 2.3** Diagram of a detector module where crystals are positioned according to the axial direction with a photodetector at both ends of each crystal element

Since this pioneer paper, several groups have used this geometrical concept where the DOI information gives the position in the axial direction [47–49]. The hit crystal gives the radial coordinates, and the axial coordinate is computed based on the light sharing on both photodetectors. The axial resolution mainly depends on the geometrical and intrinsic properties of the crystal, its coating, and the timing resolution between both sides. To achieve a reasonable axial resolution, the axial extent of the crystals needs to be adjusted, possibly limiting the axial field of view (FOV) of the PET system [50]. To overcome this issue, one solution is to add several modules in the axial direction. The challenge is then to reduce the dead space introduced by the photodetector thickness.

Another approach has been investigated to retrieve the z-coordinate of such axially oriented crystals: The AX-PET collaboration [51] developed a system with axially oriented LYSO crystal bars with wavelength shifter (WLS) strips placed orthogonally in between the layers of crystals. Photodetectors located at the end of the crystal bars and at the WLS strips are used to measure the energy deposited in both the bars and the WLS strips. The radial coordinates of each hit LYSO crystal are given by its relative position, and the z-coordinate is given by the light spread on the WLS strips. In general, these proposed geometries make it possible to obtain 3D information of the photon interactions inside the detection modules with the DOI as a discrete measurement and the axial position as a continuous one. Introducing TOF information and digital SiPM improves the resolution in the axial direction as described in [52].

To push forward toward a 3D continuous sampling of the event interaction, the use of monolithic crystals read by multichannel photodetectors has been proposed and experimentally used. This solution can improve the filling fraction with a less complex implementation of the scintillation stage. In addition, both events' positioning characteristics and DOI can be obtained through different methods, such as statistic-based positioning or maximum likelihood algorithms [53–55].

Figure 2.4 illustrates the detection strategies, going from the identification of the crystal where the interaction occurred to the localization of the interaction within a monolithic crystal.

The last decade has thus seen improvements in both hardware (photodetector) and software (optical information processing) dedicated to the 3D positioning of

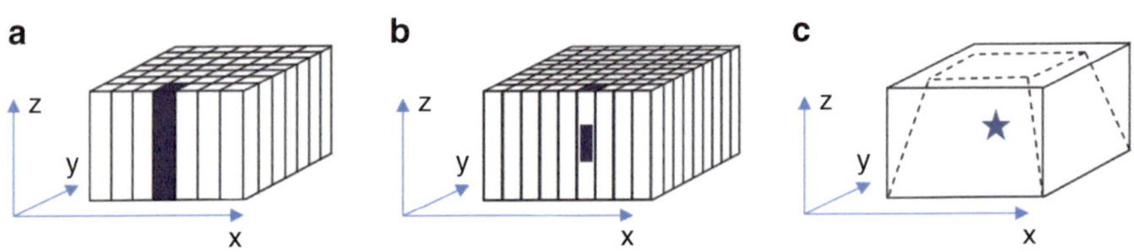


**Fig. 2.4** Schematic representation of the event positioning. (**a**) illustrates the identification of the crystal where the interaction occurred (2D), (**b**) the addition of the DOI information to the crystal identification (2.5D), and (**c**) the determination of the interaction position within a monolithic crystal (3D)

events in a monolithic crystal. In 2002, the University of Washington developed the cMICE detector [56]. An expectation-maximization (EM) algorithm, associated to a neural network, was used to read the signals. Both simulated and experimental data reported a 1.4 mm FWHM spatial resolution for a 4-mm-thick LSO crystal read by a PS-PMT. A few years later, the same team assessed the DOI resolution in an 8-mm-thick LSO crystal coupled to a H8500 MA-PMT (Hamamatsu Photonics) [57]. Both simulated and experimental data reported a 1.3 mm FWHM in the center and 1.7 mm FWHM at the edge. A parametric approach enabled to estimate the position of the scintillation leading to a spatial resolution of 1.06 mm (center) and 1.27 mm (edge) with a DOI resolution of 3.24 mm [58]. They also published the work on modeling the light response that improves the spatial resolution [59]. The latest results released have shown 1.33 ± 0.31 mm FWHM with 3.5 ± 0.31 mm DOI FWHM for an 8-mm-thick LYSO crystal with beveled edges [60].

At the same time, the University of Delft investigated the reading of a monolithic crystal using avalanche photodiodes [61]. The first results showed that the crystal surface properties have little influence on the measured spatial resolution. However, reading the optical distribution in front of the crystal allows improving the spatial resolution. In addition, the reading of the optical signal on both sides of the crystal enables to increase its thickness without deteriorating the spatial resolution. In 2007, they showed that 1.7 and 2.0 mm FWHM spatial resolution may be obtained using LSO crystals with thicknesses of 10 and 20 mm, respectively [62]. In 2009, the same team obtained results with a monolithic LYSO crystal read on each side by two APD matrices [53, 61], [63–65]. This double readout provided a spatial resolution of 1.05 mm in the center of the detector with an energy resolution of 11% and a time resolution of 2.8 ns. To improve temporal resolution, SiPM matrices were preferred [64], which led to a spatial resolution of 1.6 mm with an energy resolution of 14% and a time resolution of 960 ps.

Two preliminary papers introduced the impact of digital SiPM matrices on intrinsic detector performance. A coincidence timing resolution (CTR) of 157 ps FWHM was obtained with a 10-mm-thick LSO crystal [66] by considering the time of arrival of optical photons in the estimation algorithm. Equivalent timing results have been obtained with a 24 mm x 24 mm x 10-mm-thick LSO crystal while keeping a spatial resolution of 1 mm and an energy resolution of 12.8% [67].

The use of monolithic scintillator detectors ($32 \times 32 \times 22$ mm$^3$) with dual-sided digital SiPM readout is under investigation for high-resolution pediatric PET. The detector presents a FWHM spatial resolution of 1.1 mm, a DOI resolution of 2.4 mm, an energy resolution of 10.2%, and a coincidence resolving time of 147 ps [68].

In 2007, Benlloch et al. published the results of a small-animal PET based on LYSO continuous crystals [69]. The scintillating detector with a trapezoidal shape and a thickness of 10 mm is coupled to a H8500 MA-PMT. The 64 PMT anodes are associated to two resistive networks leading to five output signals to determine the scintillation properties and one additional signal for the trigger. This experimental setup led to a 2 mm spatial resolution at the center with an energy resolution of 19% at 511 keV.

In 2010, the same university in collaboration with the University of Pisa, thc Bruno Kessler Foundation (FBK, Trento, Italy), and the Linear Accelerator Laboratory (LAL, Orsay, France) published results obtained using SiPM matrix [70]. A 12 mm × 12 mm × 5-mm-thick LYSO crystal, with a white and black coating, was coupled to 64 channels SiPM 1.5 × 1.5 $mm^2$ read by MAROC2 ASIC-based readout board, resulting in a 0.9 mm FWHM spatial resolution and 16% energy resolution. The Valencia team designed a small-animal PET based on eight modules of MPPCs and 10-mm-thick monolithic LYSO crystals. Each MPPC array is coupled to a resistive readout circuit providing signal outputs for each row and column of the array. This configuration achieved an intrinsic detector resolution of 1.1 mm with an energy resolution of 14% [71].

In 2011, a Japanese team obtained 3 mm spatial resolution with a 15-mm-thick LYSO crystal read by a H8500 MA-PMT [72].

After almost 15 years of research programs, the use of monolithic crystals enables to reach spatial resolutions of about 1 mm comparable to or better than traditional crystal matrix solutions. Commercial preclinical PET/CT systems already offer both detection configurations.

Table 2.3 presents the data of five commercially available preclinical PET/CT systems. Additional performance on other preclinical PET systems evaluated following the NEMA NU 4–2008 can be found in [9].

**Table 2.3** Performance characteristics according to NEMA NU-4 protocol for five commercially available preclinical PET/CT scanners

| Company | Bruker | Mediso | Inviscan | Trifoil | Sedecal |
|---|---|---|---|---|---|
| Scanners | Albira | NanoPET | IRIS | LabPET12 | Argus |
| Axial FOV (mm) | 148 | 94.8 | 95 | 114 | 48 |
| Radial FOV (mm) | 80 | 123 | 80 | 100 | 67 |
| Scintillator | | | | | |
| *Size* ($mm^3$) | 50 × 50 × 10 | 1.12 × 1.12 × 13 | 1.6 × 1.6 × 12 | 2 × 2 × (14 + 12) | 1.45 × 1.45 × (8 + 7) |
| *Type* | LYSO | LYSO | LYSO | LGSO+LYSO | GSO + LYSO |
| *Shape* | Monolithic | Matrix | Matrix | Matrix, dual layers | Matrix, dual layers |
| Photodetector | MA-PMT | MA-PMT | MA-PMT | APD | MA-PMT |
| Sensitivity at center | 6.3% (358–664 keV) | 7.7% (250–750 keV) | 8% (250–750 keV) | 4.3% (250–650 keV) | 4% (250–700 keV) |
| Resolution at 5 mm | ML-EM | 2D FBP | ML-EM | 2D-FBP | 2D FBP |
| *Radial (mm)* | 1.5 | 1.7 | 1.0 | 1.5 | 1.6 |
| *Tangential (mm)* | 1.5 | 1.5 | 1.0 | 1.6 | 1.6 |
| *Axial (mm)* | 1.5 | 1.4 | 1.2 | 1.4 | 1.5 |
| Data from | [73] | [74] | [75] | [76] | [77] |

### 2.2.3 Readout Strategies

In order to extend the intrinsic performance achieved at the level of one detection module to the entire PET system, dedicated electronics is required. The quality of signal processing for nuclear coincidence detection directly affects the ability of PET scanners to extract relevant information such as energy, time stamp, and crystal identification. The data acquisition (DAQ) system therefore is one of the key components of such an instrument. The major function of the DAQ is to process and digitize the analog signals provided by the photodetectors. Most DAQ systems are built around analog subsystems to extract basic information from the detected events [78–81]. Since the current is generated from a detector, two main parameters can be extracted from the analog signal. The first one is a value directly correlated to the energy deposited in the crystal. It is often the charge given by the peak value of the integrated/shaped output signal. The second is the arrival time of the detected photon. To date, readout electronics achieve coincidence time resolutions of 120 to 335 ps [82–86]. A discriminator together with time measurement circuits such as time-to-digital converters is commonly used. The analog approach is often complex and developed for a specific and optimized detector solution but requires low power consumption [78].

With the development of Very-Large-Scale Integration (VLSI) and computer science, front-end electronics enter the era of "go digital as soon as possible." One option to fulfill this requirement is digitizing in real time the entire signal using free-running ADCs to retrieve the required timing and energy information [87]. This method requires sampling rates in the range of GHz and then could become expensive, especially if a large number of acquisition channels are needed. Another approach can be to digitize the signal already at the photodetector level using, for example, digital SiPM.

With the quest to improve as much as possible the spatial resolution, the number of electrical available signals and the complexity of the DAQ increase. One way to reduce the number of readout channels of a PET system is the use of a resistive network directly attached to the pixelated photodetector. In 1996, Siegel et al. described three different position encoding readout circuits to reduce the number of processed signals from 64 to 4 [88]. The first one was based on the scintillation camera readout approach developed by Hal Anger [13], the second was a discretized single-wire, position-sensitive proportional counter readout (DPC), and the third one was a hybrid circuit of the Anger and the DPC. These 2D readout schemes allowed reducing the number of channels being read out while preserving the detector identification accuracy. In 2003, Popov et al. proposed a novel resistive network approach called symmetric charge division (SCD) circuit [89]. This scheme has the particularity that each network input is connected by a resistance to a line and by another resistance to a column. In this case, the network makes it possible to reduce the number of channels from $n^2$ to 2n readout channels. Using this resistive network approach combined to a dedicated reconstruction algorithm, the spatial distribution of the charges on the photodetector surface can be reconstructed. The reconstructed

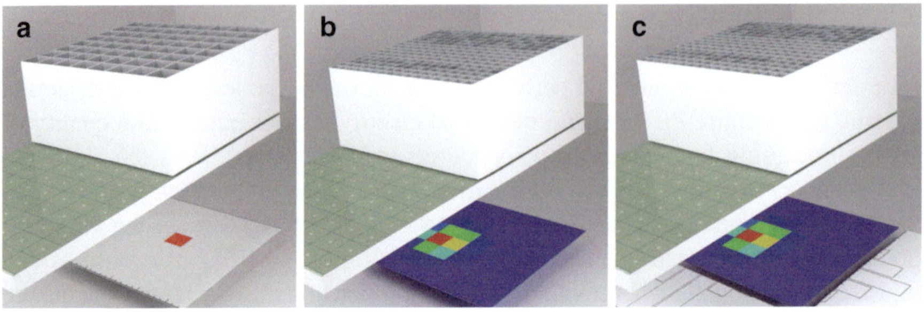


**Fig. 2.5** 3D rendering of three readout strategies. Illustrations of the charge distribution without light sharing (**a**), with light sharing (**b**), and the corresponding charge projections using a SCD readout approach (**c**)

charge distribution and the network properties provide new opportunities in terms of photodetector gain correction and depth of interaction estimation when using monolithic crystal [90]. Figure 2.5 illustrates the different strategies to extract the event positioning when using a crystal matrix.

### 2.2.4 Focus on MRI-Driven Developments

While the above section presents the recent technological advances in improving 3D location of the photon interaction and its timing, the following section shall highlight the instrumentation developments aiming at combining dedicated PET with MRI. The challenge to integrate the PET component inside a MRI system is responsible for almost all current research in instrumentation dedicated to preclinical PET imaging.

The authors would like to point readers interested in PET/MRI technology to the review recently published by Vandenberghe and Marsden, presenting a comprehensive review of this dual-modality imaging technique [91]. In this review, the authors report on all the technical challenges that have to be solved when designing (S) PE(C)T/MRI systems using SiPMs, such as temperature dependence, noise, power consumption, or cooling systems.

In 1997, a collaboration between the Crump Institute, UCLA, and Guy's and St Thomas' Clinical PET Centre in London obtained the first simultaneous PET/MRI experimental results [92, 93]. The first prototypes were based on $2 \times 2 \times 5(10)$ mm$^3$ LSO crystals coupled to PMTs via optical fibers. The performances measured in a 0.2 T magnetic field were 2 mm spatial resolution, 41–45% energy resolution, and a CTR of 20 ns FWHM. In 1998, Pichler et al. proposed another approach using LSO crystals coupled to APDs to test their performance in a tomograph prototype. $3.7 \times 3.7 \times 12$ mm$^3$ crystals were one-to-one coupled to a two-dimensional APD array. They obtained an energy resolution of 14.7%, an intrinsic spatial resolution of 2.2 mm, and a CTR of 3.2 ns FWHM [94]. Few years later, the collaboration continued the work using the same detection module and presented a spatial resolution of 2.3 mm within the whole reconstructed field of view, with a system sensitivity of 350 cps/MBq [95].

In 2006, Grazioso et al. presented an APD-based PET detector for simultaneous PET/MRI acquisition [96]. Each detector block consisted in array of APDs coupled to 2 × 2 × 20 mm$^3$ LSO crystals. Results showed an average crystal time resolution of 1.8 ns, while the average crystal energy resolution was 17%. Although no spatial resolution measurements were performed, PET and MR images of a micro-Derenzo phantom were acquired simultaneously. The 2 mm rods were the smallest holes clearly visible in the PET image.

For this period, all these research works highlighted the benefits of using APDs to design and build PET inserts compatible with MRI. The APDs' compactness as semiconductor detectors made them very suitable photodetectors and attracted the interest of the scientific community [14, 97–102].

With the introduction of SiPM, the use and interest for APD photodetector were reduced. In 2011, Kwon et al. developed a small-animal PET prototype using SiPMs [103]. The detection modules consisted of 4 × 13 arrays of 1.5 × 1.5 × 7 mm$^3$ LGSO crystals and 2 × 6 arrays of SiPMs (MPSS by Photonics SA). Performances obtained were 1.0 to 1.4 mm spatial resolution depending on the radial offset, 0.085% sensitivity for a 250–750 keV energy window, and an energy resolution of 25.8%. In 2012, Yoon et al. created a second version of the PET system developed by Kwon et al. using a similar setup with 8 × 8 SiPM arrays (Hamamatsu Photonics) coupled to 20 × 18 LGSO of 1.5 × 1.5 × 7 mm$^3$ [104]. This setup improved the sensitivity up to 0.195% for a 250–750 keV energy window. They obtained radial, tangential, and axial resolutions of 1.0, 1.2, and 1.5 mm, respectively.

In 2013, Ko et al. reported their developments of SiPM-based preclinical PET system to be inserted inside a small bore diameter and high magnetic field (>7 T) [105]. The detection modules consisted of 9 × 9 arrays of 1.2 × 1.2 × 10 mm$^3$ LYSO crystals and monolithic 4 × 4 arrays of SiPMs (Hamamatsu Photonics). Intrinsic performances obtained for the module were 0.88 to 1.09 mm spatial resolution and an energy resolution of 14.2%. At the system level, the absolute peak sensitivity was measured 3.4% for an energy window of 250–750 keV, an axial FOV of 55 mm, and a ring diameter of 64 mm [106, 107].

In 2011, Yamamoto et al. developed a SiPM-based PET system for small animal using 16 detector blocks consisting of 4 × 4 SiPM arrays (Hamamatsu Photonics) [108–110]. Two types of LGSO scintillators with different percentages of cerium doping and different lengths were used to form a DOI-encoding detector. The crystal sizes were 1.1 × 1.2 × 5 mm$^3$ and 1.1 × 1.2 × 6 mm$^3$, respectively. The measured system performances were 1.6 mm spatial resolution and 0.6% sensitivity with an energy resolution ranging from 14 to 55%. In 2014, Weissler et al. proposed a PET ring that consists of ten RF shielded detection modules [111]. Each module contains 1.3 × 1.3 × 10 mm$^3$ LYSO crystals. A 1.5-mm-thick glass plate was used to spread the light over 16 monolithic mounted and bonded SiPM arrays (FBK, Trento, Italy), resulting in a fill factor close to 92%. Performances reported were 29.7% energy resolution, 2.5 ns timing resolution, and a volumetric spatial resolution lower than 1.8 mm$^3$ in the center of the FOV.

The challenge in developing PET/MRI systems is to ensure that the overall performance is not hampered compared to stand-alone PET or MRI modalities. Eventually the PET components should be completely inserted into the open bore of the MRI gantry, between the gradient system and the MRI coil. In that configuration, restrictions regarding the PET detector size and the associated DAQ must be considered.

With the introduction of the digital version of the SiPM and the entire digitization process taken place inside the MR bore, the questions of digital electromagnetic noise patterns in the MR image were highlighted, and proper PET design shielding has been investigated [112]. The Hyperion-II$^D$ PET system was inserted in a 3 T clinical MRI system. The authors demonstrated a clear influence on the MRI environment such as a distortion of the $B_0$ field. Concerning the PET performance, the values of energy and timing resolutions are increased up to 14% [113]. A full description of this preclinical PET system and the associated performance can be found in [114].

To achieve a high level of integration, one solution is to integrate the MRI coil into the PET detector system [115]. In this published approach, the MRI coil is positioned between the scintillator and the photosensor. In that case, the material used must have optimal protection for the electromagnetic interferences, should limit the noise introduced by the PET component, and have to be transparent to the scintillation light.

The use of PET systems side-by-side with standard MRI systems already allows for a dual-modal PET/MRI approach to be performed sequentially. Moreover, several systems of PET inserts derived from academic developments (Table 2.4) and dedicated simultaneous PET/MRI systems are already available on the market.

**Table 2.4** Summary of preclinical PET detector developments compatible with MRI

| | PET module | | | | Performance | |
|---|---|---|---|---|---|---|
| | Crystal | | | | Spatial resolution (mm) | Energy resolution (%) |
| Consortium | Section (mm) | Thickness (mm) | Type | Photodetector | | |
| Shao et al. (1997) | 2 | 5 | LSO | PMT + optical fibers | 2 | 41–45 |
| Pichler et al. (1998) | 3.7 | 12 | LSO | APD | 2.2 | 14.7 |
| Ziegler et al. (2001) | 3.7 | 12 | LSO | APD | 2.3 | 22 |
| Grazioso et al. (2006) | 2 | 20 | LSO | APD | ~2.0 | 17 |
| Yamamoto et al. (2010) | 1.1 | 5 + 6 | LGSO | SiPM | 1.6 | 27 |
| Kwon et al. (2011) | 1.5 | 7 | LGSO | SiPM | 1–1.4 | 25.8 |
| Yoon et al. (2012) | 1.5 | 7 | LGSO | SiPM | 1–1.5 | 13.9 |
| Weissler et al. (2014) | 1.3 | 10 | LYSO | SiPM | 1.8 | 29.7 |
| Ko et al. (2016) | 1.2 | 10 | LYSO | SiPM | 0.7–1.3 | 14.2 |
| Schug et al. (2016) | 0.93 | 12 | LYSO | Digital SiPM | 0.9 | 12.7 |

### 2.2.5 Performance Evaluation for Preclinical PET

The outline of the possible applications for small-animal in vivo imaging is mainly driven by the performance of preclinical PET systems.

In the previous sections, we have summarized the variety of design of PET modules to address high intrinsic spatial resolution and high detection efficiency. The overall image quality also depends on the scanner geometry, the animal to be imaged, and the software suite used to reconstruct and correct the raw data.

In order to fairly compare the different existing preclinical PET systems, the National Electrical Manufacturers Association (NEMA) set out a standardized methodology for evaluating the performance of preclinical PET scanners (NEMA NU-4). The challenge for such standardized methodology is to be applicable to a large range of scanner designs with the objective to establish a baseline of system performance in typical imaging condition.

Some conditions are mandatory to use the proposed standard to evaluate the PET component: to have access to transverse sinograms and axial slices reconstructed with filtered back projection algorithm and to have a scanner transverse field of view higher than 33.5 mm in diameter to acquire data from an image quality phantom. The resulting standardized measurements can be used, for example, by manufacturers to promote their equipment and by potential customers to compare the different PET systems or to use them as gold standards for acceptance testing.

Spatial resolution, scatter fraction, noise equivalent count rate, random coincidence measurements, sensitivity, image quality, and accuracy of corrections are the main figure of merits used to evaluate the preclinical PET imaging systems.

Table 2.5 summarizes the range of performance obtained for selected commercial preclinical PET systems using the NEMA NU-4 standards. The purpose of this table is not to list, in an exhaustive manner, all the performance of the entire set of preclinical PET systems but to give to the reader a range of performance achievable with existing commercial systems.

## 2.3 Instrumentation in SPECT

Like PET imaging, single-photon emission computed tomography (SPECT) is based on the tracer principle. Although using a similar detector technology to that of PET, the use of radioactive isotopes emitting simple photons requires the use of a collimator to determine the origin of gamma photons. Thus, developments relating to SPECT systems are essentially based on the collimation stage that is most likely to fulfill the specifications of the instrument. The following sections present the recent instrumental developments dedicated to preclinical SPECT imaging, with an emphasis on SPECT systems dedicated to combined SPECT/MRI. We will also discuss the main collimation profiles and the geometric parameters responsible for

**Table 2.5** Non-exhaustive range of performance obtained for selected commercial preclinical PET systems based on the NEMA NU-4 evaluation procedure

| **System characteristic** | | |
|---|---|---|
| Ring diameter | 110–260 mm | |
| Axial FOV | 45–130 mm | |
| Crystal section | 1.1–2.2 mm | |
| Crystal thickness | 10–25 mm | |
| **System performance** | | |
| Spatial resolution (FWHM) | At 5 mm | At 25 mm |
| Transverse | 1–2.3 mm | 1.4–2.9 mm |
| Axial | 1–2.3 mm | 1.4–3.3 mm |
| Peak detection efficiency | 1.2–8% | |
| Scatter fraction | | |
| Mouse | 5–30% | |
| Rat | 12–35% | |
| **Image quality** | | |
| Uniformity | 4.5–15.5% | |
| Recovery coefficient for 1 mm | 3–27% | |
| Recovery coefficient for 5 mm | 75–100% | |

the performance of the system, before presenting alternate collimation approaches, which open new perspectives.

### 2.3.1 Overview of Instrumentation Advances

In 2006, Ronald J. Jaszczak published a wonderful paper describing the origins of SPECT imaging, which took place about 60 years ago [116]. This historical review mentions the beginnings of this imaging modality through stories from pioneers who shaped the SPECT instrumentation. Although based on concepts common to those offered by these pioneers, SPECT instrumentation remains a research field of interest. Many SPECT systems, especially preclinical systems, have emerged over the last 20 years. Technological advances have enabled and constantly help to improve the overall performance of these systems. Table 2.6 presents the basic performance of noncommercial preclinical SPECT scanners developed over the past decade [117–123].

These academic developments aimed to explore all the possibilities offered by the different modes of detection and collimation profiles. In parallel with these developments, manufacturers have also proposed preclinical systems. It is interesting to note that these commercial systems are all based on a crystal-PMT approach associated with a (multi-) pinhole collimation. The use of large detection FOVs and strong magnification factors allow them in particular to increase the detection efficiency with very high spatial resolutions. Table 2.7 presents some of the latest commercial SPECT/CT systems for preclinical investigations.

**Table 2.6** Performance characteristics of selected noncommercial preclinical SPECT scanners

| System | Spatial resolution (best) | Sensitivity (best) | Detector | Collimation |
|---|---|---|---|---|
| HiRe-SPECT | 1.35 mm | 0.0035% | CsI(Na) + PMTs | Slit |
| Fast-SPECT II | 0.65 mm | 0.0400% | NaI(Tl) + PMTs | Pinholes |
| LOrA-SPECT | 0.75 mm | 0.1100% | CsI(Na) + PMTs | Slit |
| CoALA-SPECT | 1.20 mm | 0.0150% | NaI(Tl) + PMTs | Parallel-hole |
| A-SPECT | 0.74 mm | 0.0180% | NaI(Tl) + PMTs | Pinholes |
| Semi-SPECT | 1.45 mm | 0.0077% | CdZnTe + Asic | Pinholes |
| Mouse-SPECT | 1.70 mm | 0.0029% | NaI(Tl) + PMTs | Pinholes |

**Table 2.7** Performance characteristics of preclinical SPECT/CT scanners from three major manufacturers

| Company scanners | Siemens Healthcare Inveon SPECT | Mediso NanoSPECT | MILabs U-SPECT+ |
|---|---|---|---|
| Maximum FOV | 110 × 130 mm | 200 × 250 mm | NC |
| Scintillator | NaI (2 × 2 × 10) | NaI (monolithic) | NaI (monolithic) |
| Sensitivity (max) | 0.04% | 1.30% | 1.00% |
| Resolution at CFOV (best) | 0.60 mm | 0.30 mm | 0.25 mm |

### 2.3.2 Scintillation Versus Direct Conversion

Over the past decades, thallium-doped Sodium Iodide has been the scintillating crystal of choice when building SPECT imaging systems. Based on the Anger camera, which consists in a monolithic crystal coupled to photomultiplier tubes, SPECT detector heads have presented an intrinsic spatial resolution of about 3 mm. The reliability, robustness, and relatively low cost of these detectors are the reasons why they are still used today in preclinical systems. However, their intrinsic spatial resolution requires the use of collimators presenting a magnification factor in order to obtain spatial resolutions that are suitable for imaging the small animal.

The quest for an ever better intrinsic spatial resolution has led to the use of crystal matrices. Pixelated crystal arrays comprising small tightly packed and optically isolated crystals are now used in preclinical systems to ensure high intrinsic spatial resolution. The intrinsic spatial resolution of the detector is approximately assume to be the same as the crystal pitch, provided it is coupled to a high-resolution detector capable of resolving individual crystal elements, such as a PS-PMT or SiPM. While a small crystal pitch is desirable for high spatial resolution, the thickness must be sufficient to absorb most photons for good detection efficiency. However, as the crystal length-to-width ratio increases, the light output decreases due to internal reflections, which leads to an energy resolution loss [124–126].

Several inorganic scintillators were used in preclinical systems. Table 2.8 lists the principal properties of thallium-doped Sodium Iodide (NaI[Ce] or NaI), thallium-doped Cesium Iodide (CsI[Ce] or CsI), cerium-doped Yttrium Aluminium Perovskite ($YAlO_3$[Ce] or YAP), and cerium-doped Lanthanum Bromide ($LaBr_3$[Ce] or $LaBr_3$), cerium-doped Lanthanum Chloride ($LaCl_3$[Ce] or $LaCl_3$) crystals.

**Table 2.8** Properties of scintillating crystals commonly used in single-photon imaging

| Scintillation crystal | NaI | CsI | YAP | $LaBr_3$ | $LaCl_3$ |
|---|---|---|---|---|---|
| Density (g $cm^{-1}$) | 3.67 | 4.51 | 5.55 | 5.29 | 3.79 |
| Index of refraction | 1.85 | 1.79 | 1.94 | 1.90 | 1.90 |
| Luminescence (ph $keV^{-1}$) | 38 | 54 | 18 | 63 | 49 |
| Peak emission wavelength (nm) | 415 | 550 | 350 | 380 | 350 |
| Emission decay time (ns) | 250 | 1000 | 27 | 26 | 28 |
| Attenuation length at 140 keV (cm) | 0.41 | 0.28 | 0.58 | 0.29 | 0.37 |
| Hygroscopicity | Yes | Poor | No | Yes | Yes |

Although the use of a crystal-photodetector pair remains the main choice for detecting gamma rays, semiconductor detectors based on CdTe or CdZnTe were used to directly convert gamma rays into electrical charges. With respect to scintillation devices, solid-state detectors provide a larger conversion yield: typically 30,000 charges for a 140 keV energy release. As a consequence, the energy resolution is not limited by charge generation statistics but by other phenomena like electrical noise or material uniformity. To date, the best devices achieve an energy resolution below 2% at 140 keV, while standard systems are close to 5% at 140 keV [127–129]. This opens a new path to properly investigate the development of dual-isotope imaging protocols [130].

Another advantage of semiconductor detectors is their extremely good spatial resolution, which is not limited by light spreading and photon statistics but rather by the readout circuitry. A typical device resolution is of the order of 2.5 mm, but the use of high density readout [129] or sub-pixel positioning electronics [131] allows to obtain an intrinsic resolution of few hundreds of micrometers (200–400 μm). Additionally, CdTe- or CdZnTe-based detectors are nowadays integrated in small modules that couple the semiconductor crystals and the readout electronics on the same substrate. This module compactness presents an interesting way to build innovative SPECT systems that enhance sensitivity and resolution.

### 2.3.3 Focus on Collimation

For over 50 years, several collimators have been proposed. The most known and used are the parallel-hole collimator [132–134], the (multi-)pinhole collimator [135–139], and the fan-beam/cone-beam collimator [140–143]. Although less used or still the subject of current research, other collimators have been proposed, such as coded aperture [120, 144] or slit-slat collimators [145–147]. All these collimators present different resolution-sensitivity trade-offs, which are directly related to their geometries. Mainly governed by collimation parameters, the resolution-sensitivity trade-off is one of the main factors determining the collimator the most suitable for an intended study. It is therefore essential to optimize these parameters to get the best system performance. Several approaches have been proposed to optimize the collimation parameters. The most commonly used method remains the use of Monte Carlo methods. For instance, in 2004, Song et al. investigated the optimal pinhole diameter and channel height using Monte Carlo simulations. Trade-off curves of spatial resolution and sensitivity were obtained for pinhole diameters varying from 0.25 mm to 2 mm and channel heights varying from 0 to 1 mm (knife-edge) [148]. Similar works have been conducted on parallel-hole collimators. Monte Carlo simulations have been used to optimize the hole diameter, the collimation height, the septal thickness, and the collimation material [141, 148, 149].

Although these collimators enable a large number of studies, a low sensitivity can limit the use of single-photon imaging and SPECT systems. One alternative to the parallel-hole or the pinhole approaches is the slit-slat collimation [145–147]. The slit-slat profile combines fan-beam and parallel-beam properties and allows a better detection efficiency than the multi-pinhole collimation [147]. Similar to the pinhole and cone-beam collimation, the slit-slat approach presents an improved transverse resolution provided by a transaxial magnification. Furthermore, the axial septa lead to an extended axial FOV compared to pinhole imaging [146]. Figure 2.6 presents the common collimation profiles and their corresponding parameters.

A second alternative is the rotating slat collimator (RSC). Originally, the design of a linear detector was proposed independently by Keyes (1975) and Tosswill (1977) [150, 151]. In 1979, Entine et al. combined a CdTe detector with a parallel plate collimator [152]. In 2001, Gagnon et al. used RSC in combination with solid-state detectors in the SOLSTICE (SOLid STate Imager with Compact Electronics) system [153]. This collimation profile provides a better trade-off between spatial resolution and detection efficiency than a parallel-hole collimator [153, 154]. In addition, Webb (1993) previously showed that the use of a RSC combined with a conventional SPECT detector leads to a sensitivity enhancement of 40 times [155]. Although the pinhole approach provides a good trade-off

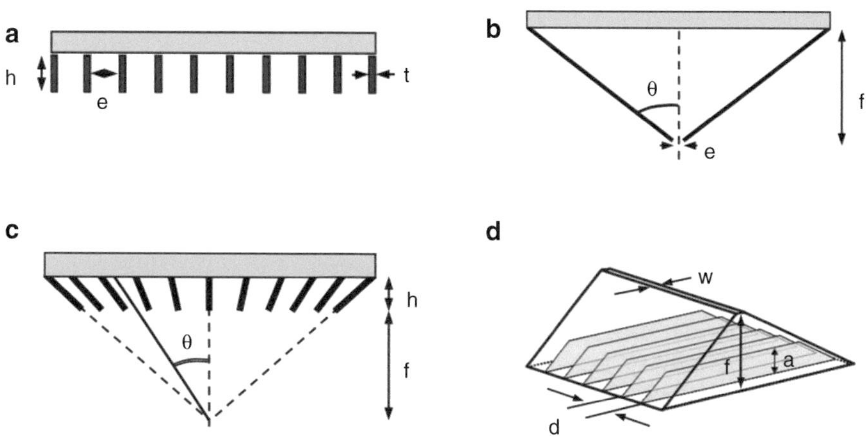


**Fig. 2.6** Schematics of four different collimation profiles. (**a**) Parallel-hole collimator. e corresponds to the aperture dimension, h the collimation height, and t the septal thickness. (**b**) (Single) pinhole collimator. f corresponds to the focal length and θ the opening angle. (**c**) Fan-beam collimator and (**d**) slit-slat profile. d corresponds to the gap between two consecutive slats, a is the slat height, and W the slit width

between spatial resolution and sensitivity, the price in most cases is a reduction in the reconstructed field of view where RSC allows for a similar FOV to be maintained [136]. RSCs are fundamentally different from other collimators. RSCs do not measure line integrals but plane integrals acquired over an extra rotation of the detector around its own axis, called spin rotation. This rotation allows the complete set of one-dimensional (1D) projections to be obtained [145, 146, 156], which is then used to reconstruct two-dimensional (2D) projections. However, this necessary reconstruction step makes it difficult to directly compare RSC to other collimators in terms of sensitivity. For parallel-hole collimators, sensitivity is directly related to the collimator transparency, a relation biased by the reconstruction step in the case of RSC.

More recently, Boisson et al. proposed a dedicated detection system presenting both high sensitivity and submillimetric spatial resolution [157]. The innovative aspect of this system was the use of a YAP:Ce crystal segmented into 32 slices of 0.570 mm wide, associated with a RSC. This 1D geometry however imposes the rotation of the entire detection module (including the collimator), thus reducing the reconstructed field of view. They demonstrated later the value of using a system matrix to reconstruct 2D projections for a RSC system [158]. This work also aimed at studying the resolution-sensitivity trade-offs obtained by varying different collimation parameters: the slats height (H) and the gap between two consecutive slats (g). The GATE (Geant4 Application for Tomographic Emission, [159]) simulation platform was used to generate probability matrices corresponding to (H, g) couples offering the best sensitivity-spatial resolution trade-off for specific applications. Based on these preliminary simulation results, they showed that both a submillimeter spatial resolution and sensitivity greater than 0.2% could be obtained using optimized collimation parameters considering a submillimetric intrinsic spatial resolution [160]. In the context of a full SPECT system, such a preclinical prototype opens new perspectives. A SPECT system consisting in four detection heads would provide a sensitivity of about 0.8% while keeping (1) a submillimeter spatial resolution and (2) the initial $50 \times 50$ mm$^2$ FOV.

### 2.3.4 Rotating Versus Stationary SPECT Systems

Most clinical and small-animal SPECT systems consist in detector-collimator pairs mounted on a gantry that rotates around the subject considering either a precise step-and-shoot or a continuous motion. Multiple detector systems allow sensitivity improvement and reduce the need for a system to acquire projection data over a full 180° or 360°. However, small-animal SPECT systems based on a pinhole collimation approach often suffer from a limited FOV due to high magnification or small compact detectors. One option to overcome this limitation is to increase the radius of rotation, but this results, in most cases, in the reduction of both the system's spatial resolution and sensitivity. Despite these possible limitations, currently available commercial systems achieve submillimeter spatial resolution and absolute sensitivity greater than 1%.

Several systems have been developed based on another approach: stationary detectors that surround the animal and multiple stationary or rotating pinholes [118, 123, 136, 161, 162]. Based on multi-pinholes, these systems present a high sensitivity, while avoiding mechanical misalignments due to gantry rotation. But the most important advantage of stationary SPECT systems is their ability to perform dynamic SPECT of tracers with fast kinetics [118, 162] due to the simultaneous acquisition of the entire set of projections, which eliminates potential reconstruction errors due to redistribution of the radiopharmaceutical.

### 2.3.5 Multimodality Systems

It is becoming increasingly important in research applications to accurately localize radiopharmaceutical biodistribution relative to known anatomical structures. Both X-ray CT and MR provide detailed anatomical information, which is highly complementary to the SPECT study. Thus, it is common in commercial systems for PET or SPECT to be one component of a dual- or trimodality imaging system. There are several possible combinations and approaches to multimodality imaging, including PET/CT, SPECT/CT, PET/MR, and SPECT/MR. In the following section, we will focus on the latest multimodal developments: SPECT/MR imaging systems.

Like instrumental developments dedicated to PET/MRI, recent years have seen several projects to develop SPECT inserts that can be integrated into an MRI. Although SPECT/CT systems remain the reference in a multimodal approach for single-photon imaging, combining SPECT and MRI information opens new perspectives.

Some technical challenges inherent in the development of SPECT and PET inserts are similar, namely, the constraints of size or the susceptibility of detectors in the magnetic field. However, in the case of SPECT, an essential element adds to the complexity of development: the collimator. The need to use a collimator adds to the constraint of space and thereby reduces the useful field of view of the insert. It is also essential to design a collimator in a material, which does not disturb the magnetic field of the MRI, but with a geometry that would provide acceptable performance. Typically, a SPECT insert should present a spatial resolution of one millimeter or less and the greatest efficiency possible. Here we only present recent instrumental developments of SPECT inserts dedicated to simultaneous SPECT/MRI. We invite readers to read a full review published by Van Audenhaege et al. (2015) concerning the selection, optimization, and manufacturing methods of collimators within the framework of clinical and preclinical imaging and mentioning the SPECT/MRI approach [163]. Although SPECT inserts are still in early stages of development, the following paragraph presents the latest research developments in SPECT instrumentation dedicated to simultaneous SPECT/MRI and their performance.

The benefits of combining SPECT and MRI information are undeniable. However, the concept of simultaneous SPECT/MRI is relatively recent, with the first prototypes appearing in the late 2000s. In 2009, Wagenaar et al. developed a stationary SPECT system for simultaneous preclinical SPECT/MRI based on CZT. The system consists in four rings of eight CZT modules, 16 × 16 pixels of 1.6 mm each. They considered a cylindrical arrangement of pinholes with a magnification of approximately 1.0. This stationary tomographic SPECT system is capable of obtaining 24 views simultaneously. Two different arrangements of pinholes are used to perform whole-body or organ-specific acquisitions. A spatial resolution of about 1.5 mm FWHM with the 1.6 mm pixels and a pinhole collimator of 1.0 mm diameter was reported. Limiting factors such as a limited volume within the magnet bore and a magnification of 1.0 are mostly responsible for this loss in terms of spatial resolution. In 2011, Meier et al. continued this work using the same CZT detector modules [164]. They reported an average energy resolution of 5.4 keV FWHM at 122 keV. Sensitivity and spatial resolution measurements were performed considering a magnification of 0.85. They obtained a sensitivity of 0.062% using a 2 mm diameter pinhole and 2.2 mm FWHM spatial resolution for a 0.5 mm aperture. These collaborative works presented preliminary results of the first SPECT insert dedicated to simultaneous SPECT/MRI. In 2014, Cai et al. proposed an alternative design of a stationary SPECT insert based on ten energy-resolved photon-counting (ERPC) detectors assembled as a compact ring [165]. The distance between the opposite detectors is 15.6 cm, and the detection area of each detector is 22.5 × 45 $mm^2$. Each detector proposes four 300 or 500 μm pinholes, with a magnification factor of 1.2. The object-to-pinhole distance is designed to be around 36 mm. Pinhole inserts are made of cast platinum (90%)-iridium (10%) alloy, which provides the maximum stopping power and are compatible with MR scanners. A total of 40 pinholes were used in the 10-detector ring system leading to sensitivity up to 0.07%. In terms of spatial resolution, the system was able to clearly resolve the 500 mm features of a homemade Jaszczak phantom using the 300 mm pinholes.

In 2013, the INSERT research project was funded by the Seventh Framework Programme of the European Commission. The objective of the INSERT project is to develop an innovative system combining SPECT and MRI for simultaneous imaging. The system will be validated at both the preclinical and clinical levels thanks to the creation of animal models and to a pilot study, which will involve patients with glioma. In 2014, Busca et al. published the simulation results of the expected performance of INSERT for both preclinical and clinical imaging [166]. The fundamental unit is a 5 × 5 cm gamma camera, based on the Anger architecture with a continuous scintillator (e.g., CsI(Tl), NaI(Tl)), read out by an array of silicon photodetectors. Two possible solutions have been taken into account for the scintillator readout: silicon drift detectors (SDDs) and SiPMs. The basic SDD photodetector unit consists of a matrix composed of nine square SDDs, each one of 8 × 8 mm active area, in a final 3 × 3 format, for an overall size of 26.08 × 26.08 $mm^2$. In the simulation, a 8-mm-thick CsI(Tl) scintillator with parallelepiped edges was considered, providing a final UFOV of about 4 × 4 $cm^2$. The intrinsic spatial resolution (FWHM) was estimated by a Gaussian fitting of the reconstructed points with values between 0.85 mm (center) and 1.05 mm (borders). The energy resolution was found between 9.6 and 14.6% when the system is cooled down to −20 °C. The second option is based on a detection plane composed by 10 × 10 single SiPMs. To shortcut the output terminal, they considered a channel merging with a single merged unit composed by 2 × 2 SiPMs. Also in this simulation, a CsI(Tl) scintillator wrapped with Teflon and with parallelepiped edges is considered. The resulting UFOV is roughly 4 × 4 $cm^2$, and the intrinsic spatial resolution was estimated between 1.3 mm (center) and 1.5 mm (borders). The energy resolution has been evaluated with both analytical formulation and Monte Carlo simulations between 11.6 and 15.9% also in this case. The preclinical SPECT system will be composed by one circular ring of detection modules designed for MRI (7 T and 9.4 T) with 20 cm aperture. The necessary UFOV of the SPECT system for mouse brain would be 16 mm transaxially x 11 mm axially, whereas for rat brain 20 × 20 mm. Considering a FOV of the detectors larger in both transaxial and axial directions than the animal size, a multi-pinhole technology with divergent projections was chosen for the preclinical imaging system. While a 1 mm spatial resolution was measured for both configurations, effective sensitivity top values of 0.22% and 0.094% were obtained for the mouse and rat brain configurations, respectively. The aim of the clinical system is to image a whole human brain (FOV 20 × 15 cm) to avoid the need for prior knowledge of tumor location from other modalities. The SPECT system should be composed of at least two adjacent circular rings of detection modules to be inserted in a 59 cm aperture 3 T PET/MRI system (Siemens). The clinical design is oriented to maximize sensitivity with an overall resolution maintained at a level similar to that achieved with conventional gamma camera-based SPECT (8–10 mm). Higher sensitivity can be achieved with the use of slit-slat collimator. Simulations of a Derenzo-type phantom were performed considering a slit-slat collimator and a 42.4 mm projection UFOV, which led to a targeted resolution of 8 mm.

## 2.4 Summary

Whether based on academic or commercial developments, preclinical SPECT systems now offer a spatial resolution of less than 0.3 mm and a sensitivity of up to 1.3%, where preclinical PET systems' sensitivity is on the order of 10% and a spatial resolution down to 0.6 mm. Although it is fundamentally not possible to compare the performance of PET and SPECT systems, the latter are nevertheless governed by a set of intrinsic trade-offs. Preclinical PET systems are intrinsically not limited in terms of sensitivity, unlike SPECT systems whose sensitivity is strongly impacted by the collimation stage. Conversely, PET systems resolution is limited by various factors such as the positron range among others, unlike SPECT systems, which are intrinsically not limited in terms of spatial resolution. However, it should be noted that the improvement of the spatial resolution of SPECT systems almost inevitably leads to the reduction of the system's field of view.

Unlike PET systems dedicated to small-animal imaging, for which NEMA has defined the standard NU 4–2008, preclinical SPECT systems are, to date, subject to no standards. A task force has recently been assigned to establish what could be the new NEMA NU-5, also including a new phantom, adapted to small-animal SPECT systems and to define the IQ parameters.

PET and SPECT systems show constantly improving performance. These modalities should not be opposed but rather associated in a multimodal approach to open new perspectives in the field of molecular imaging.

## References

1. Massoud TF, Gambhir SS. Molecular imaging in living subjects: seeing fundamental biological processes in a new light. Genes Dev. 2003;17:545.
2. James ML, Gambhir SS. A molecular imaging primer: modalities, imaging agents, and applications. Physiol Rev. 2012;92:897.
3. Flohr TG, et al. Multi-detector row CT systems and image-reconstruction techniques. Radiology. 2005;235:756.
4. Brasse D, et al. Towards an inline reconstruction architecture for micro-CT systems. Phys Med Biol. 2005;50:5799.
5. O'Connor MK, Kemp BJ. Single-photon emission computed tomography/computed tomography: basic instrumentation and innovations. Semin Nucl Med. 2006;36:258.
6. Madsen MT. Recent advances in SPECT imaging. J Nucl Med. 2007;48:661.
7. Khalil MM, et al. Molecular SPECT imaging: an overview. Int J Mol Imaging. 2011;2011:796025.
8. Levin CS, Zaidi H. Current trends in preclinical PET system design. PET Clinics. 2007;2:125.
9. Goertzen AL, et al. NEMA NU 4-2008 comparison of preclinical PET imaging systems. J Nucl Med. 2012;53:1300.
10. Hollingworth W, et al. The diagnostic and therapeutic impact of MRI: an observational multicentre study. Clin Radiol. 2000;55:825.
11. Pichler BJ, et al. PET/MRI: paving the way for the next generation of Clinical multimodality imaging applications. J Nucl Med. 2010;51:333.
12. Ebbini ES, Ter Haar G. Ultrasound-guided therapeutic focused ultrasound: current status and future directions. Int J Hyperth. 2015;23:1.
13. Anger HO. Scintillation camera. Rev Sci Instr. 1958;29:27.
14. Pichler BJ, et al. Performance test of an LSO-APD detector in a 7-T MRI scanner for simultaneous PET/MRI. J Nucl Med. 2006;47:639.
15. Keereman V, et al. Temperature dependence of APD-based PET scanners. Med Phys. 2013;40:092506.
16. Chatziioannou AF. PET scanners dedicated to molecular imaging of small animal models. Mol Imaging Biol. 2002;4:47–63.
17. Casey ME, Nutt R. A multicrystal two dimensional BGO detector system for positron emission tomography. IEEE Trans Nucl Sci. 1986;33:460–3.
18. Conti M, et al. Characterization of $^{176}$Lu background in LSO-based PET scanners. Phys Med Biol. 2017;62:3700–11.
19. Lecoq P. Development of new scintillators for medical applications. Nucl Instrum Methods Phys Res A. 2016;809:130–9.
20. Derenzo SE, et al. The quest for the ideal inorganic scintillator. Nucl Instrum Methods Phys Res A. 2003;505:111.
21. Otte AN, et al. A test of silicon photomultipliers as readout for PET. Nucl Instrum Methods Phys Res A. 2005;545:705.
22. Ginzburg D, et al. Optimizing the design of a silicon photomultiplier-based radiation detector. Nucl Instrum Methods Phys Res A. 2011;652:474. https://doi.org/10.1016/j.nima.2011.01.022.

23. Thiessen JD, et al. Performance evaluation of SensL SiPM arrays for high-resolution PET. IEEE Nucl Sci Symp Conf Rec. 2013:M21.
24. Frach T, et al. The digital silicon photomultiplier - principle of operation and intrinsic detector performance. In: 2009 IEEE Nucl. Sci. Symp. Conf. Rec. (Orlando, FL, USA, 24 Oct.–1 Nov.); 2009, p. 1959–1965.
25. Gu Y, Levin CS. Study of electrode pattern design for a CZT-based detector. Phys Med Biol. 2014;59:2599–621.
26. Yoo, et al. Simulation for CZT Compton PET (Maximization of the efficiency for PET using Compton event). Nucl Instrum Methods Phys Res A. 2011;652:713–6.
27. Gu Y, et al. Study of a high-resolution, 3D positioning cadmium zinc telluride detector for PET. Phys Med Biol. 2011;56:1563–2011.
28. Levin CS. Promising new photon detection concepts for high-resolution clinical and preclinical PET. J Nucl Med. 2012;53:167–70.
29. Abbaszadeh S, et al. Positioning true coincidences that undergo inter- and intra-crystal scatter for a sub-mm resolution cadmium zinc telluride-based PET system. Phys Med Biol. 2018;63(2):025012. https://doi.org/10.1088/1361-6560/aa9a2b.
30. Thompson C. The effects of detector material and structure on PET spatial resolution and efficiency. IEEE Trans Nucl Sci. 1990;37:718–24.
31. Chung YH, et al. Characterization of dual layer phoswich detector performance for small animal PET using Monte Carlo simulation. Phys Med Biol. 2004;49:2881.
32. Jung JH, et al. Optimization of LSO/LuYAP phoswich detector for small animal PET. Nucl Instrum Meth Phys Res A. 2007;571:669–75.
33. Eriksson L, et al. Design Considerations of phoswich detectors for high resolution positron emission tomography. IEEE Trans Nucl Sci. 2009;56:182–8.
34. Ito M, et al. A Four-Layer DOI detector with a relative offset for use in an animal PET system. IEEE Trans Nucl Sci. 2010;57:976–81.
35. Roncali, et al. Design considerations for DOI-encoding PET detectors using phosphor-coated crystals. IEEE Trans Nucl Sci. 2014;61:67–73.
36. Shao Y, et al. Design studies of a high resolution PET detector using APD arrays. IEEE Trans Nucl Sci. 2000;47:1051–7.
37. Shao Y, et al. Dual APD array readout of LSO crystals: Optimization of crystal surface treatment. IEEE Trans Nucl Sci. 2002;49:649–54.
38. Dokhale PA, et al. Performance measurements of a depth-encoding PET detector-module based on position-sensitive avalanche photodiode readout. Phys Med Biol. 2004;49:4293.
39. Dokhale PA, et al. Intrinsic spatial resolution and parallax correction using depth-encoding PET detector modules based on position-sensitive APD readout. IEEE Trans Nucl Sci. 2006;53:2666–70.
40. James SS, et al. Experimental characterization and system simulations of depth of interaction PET detectors using 0.5 mm and 0.7 mm LSO arrays. Phys Med Biol. 2009;54:4605–19.
41. Yang Y, et al. Depth of interaction resolution measurements for a high resolution PET detector using position sensitive avalanche photodiodes. Phys Med Biol. 2006;51:2131–42.
42. Moses WW, et al. A room temperature LSO/PIN photodiode PET detector module that measures depth of interaction. IEEE Trans Nucl Sci. 1995;42:1085–9.
43. Chaudhari AJ, et al. PSPMT/APD hybrid DOI detectors for the PET component of a dedicated breast PET/CT system: A feasibility study. IEEE Trans Nucl Sci. 2008;55:853–61.
44. Godinez F, et al. Characterization of a high-resolution hybrid DOI detector for a dedicated breast PET/CT scanner. Phys Med Biol. 2012;57:3435–49.
45. Godinez F, et al. Development of an ultra high resolution PET scanner for imaging rodent paws: PawPET. IEEE Trans Rad Plas Med Sci. 2018;2:7–16.
46. Ter-Pogossian MM, et al. A multislice positron emission computed tomograph (PETT IV) yielding transverse and longitudinal images. Radiology. 1978;128:477–84.
47. Shimizu K, et al. Development of 3-D detector system for positron CT. IEEE Trans Nucl Sci. 1988;35:717–20.

48. Braem A, et al. Feasibility of a novel design of high resolution parallax-free Compton enhanced PET scanner dedicated to brain research. Phys Med Biol. 2004;49:2547.
49. Jan S, et al. GePETоS: A Geant4 Monte Carlo simulation package for positron emission tomography. IEEE Trans Nucl Sci. 2005;52:102–6.
50. Salvador S, et al. Design of a high performances small animal PET system with axial oriented crystals and DOI capability. IEEE Trans Nucl Sci. 2009;56:17–23.
51. Beltrame P, et al. The AX-PET demonstrator-Design, construction and characterization. Nucl Instrum Methods Phys Res A. 2011;654:546–59.
52. Casella C, et al. A high resolution TOF-PET concept with axial geometry and digital SiPM readout. Nucl Instrum Methods Phys Res A. 2014;736:161–8.
53. Maas MC, et al. Monolithic scintillator PET detectors with intrinsic depth of interaction correction. Phys Med Biol. 2009;54:1893–908.
54. van Dam HT, et al. Improved nearest neighbor methods for gamma photon interaction position determination in monolithic scintillator PET detectors. IEEE Trans Nucl Sci. 2011;58:2139–47.
55. van Dam HT, et al. A practical method for determining depth of interaction in monolithic PET scintillator detectors. Phys Med Biol. 2011;56:4135–45.
56. Joung J, et al. cMICE: a high resolution animal PET using continuous LSO with a statistics based positioning scheme. Nucl Instrum Methods Phys Res A. 2002;489:584–98.
57. Ling T, et al. Depth of interaction decoding of a continuous crystal detector module. Phys Med Biol. 2007;52:2213–28.
58. Ling T, et al. Parametric positioning of a continuous crystal PET detector with depth of interaction decoding. Phys Med Biol. 2008;53:1843–63.
59. Miyaoka RS, et al. Detector response modeling light for a thick slab continuous detector. J Nucl Sci Technol. 2008;45:634–8.
60. Miyaoka RS, et al. New miniature continuous crystal element (cMICE) detector geometries. IEEE Nucl Sci Symp Conf Rec. 2009;2009:3639.
61. Maas MC, et al. Experimental characterization of monolithic crystal small-animal PET detectors read out by APD arrays. IEEE Trans Nucl Sci. 2006;53:1071–7.
62. Bruyndonckx P, et al. Towards a continuous crystal APD-based PET detector design. Nucl Instrum Methods Phys Res A. 2007;571:182–6.
63. Ling T, et al. Performance comparisons of continuous Miniature Crystal elemento (cMICE) detectors. IEEE Trans Nucl Sci. 2006;53:2513–8.
64. Schaart DR, et al. A novel, SiPM-array-based monolithic scintillator detector for PET. Phys Med Biol. 2009;54:3501–12.
65. van der Lann DJ, et al. Optical simulation of monolithic scintillator detectors using GATE / Geant4. Phys Med Biol. 2010;55:1659–75.
66. van Dam HT, et al. Sub-200 ps CRT monolithic scintillator in PET detectors using digital SiPM arrays and maximum likelihood estimation time interaction. Phys Med Biol. 2013;58:3243–57.
67. Seifert S, et al. First characterization of a digital SiPM based time-of-flight PET detector with 1 mm spatial resolution. Phys Med Biol. 2013;58:3061–74.
68. Borghi G, et al. A 32 mm × 32 mm × 22 mm monolithic LYSO:Ce detector with dual-sided digital photon counter readout for ultra-performance TOF-PET and TOF-PET/MRI. Phys Med Biol. 2016;61:4929–49.
69. Benlloch JM, et al. Scanner calibration of a small animal PET Camera based on continuous LSO crystals and flat panel PSPMTs. Nucl Instrum Methods Phys Res A. 2007;571:26–9.
70. Llosa G, et al. Characterization of a PET detector head based on continuous LYSO crystals and monolithic, 64-pixel silicon photomultiplier arrays. Phys Med Biol. 2010;55:72997315.
71. Gonzales A, et al. Small animal PET based on 16x16 TSV-MPPCs and monolithic crystals. Eur J Nucl Med Mol Imaging Physics. 2015;2:A16.
72. Yoshida E, et al. Basic performance of a wide area PET detector with a monolithic scintillator. Phys Radiol Technol. 2011;4:4134–9.

73. Spinks TJ, et al. Quantitative PET and SPECT performance characteristics of the Albira Trimodal pre-clinical tomograph. Phys Med Biol. 2014;59:715–31.
74. Szanda I, et al. National Electrical Manufacturers Association NU-4 performance evaluation of the PET component of the NanoPET/CT preclinical PET/CT scanner. J Nucl Med. 2011;52:1741–7.
75. Belcari N, et al. NEMA NU-4 performance evaluation of the IRIS PET/CT preclinical scanner. IEEE Trans Rad Plas Med Sci. 2017;1:301–9.
76. Bergeron M, et al. Imaging performance of LabPET APD-based digital PET scanners for pre-clinical research. Phys Med Biol. 2014;59:661.
77. Wang, et al. Performance evaluation of the GE Healthcare eXplore VISTA dual-ring small-animal PET scanner. J Nucl Med. 2006;47:1891–900.
78. Habte F, Levin CS. Investigation of low noise, low cost readout electronics for high sensitivity PET systems based on Avalanche Photodiode arrays. IEEE Nucl Sci Symp Conf Rec. 2002;2:661.
79. Monzo JM, et al. Evaluation of a timing integrated circuit architecture for continuous crystal and SiPM based PET systems. JINST. 2012;8:C03017.
80. Espana S, et al. Performance evaluation of SiPM photodetectors for PET imaging in the presence of magnetic fields. Nucl Instrum Methods Phys Res A. 2010;613:308.
81. Zeng H, et al. Design of a novel front-end readout ASIC for pet imaging system. J Signal Inform Proc. 2013;4:129.
82. Dey S, et al. Impact of analog IC impairments in SiPM interface electronics. IEEE Trans Nucl Sci. 2012;2012:3572.
83. Jarron P, et al. Time based readout of a silicon photomultiplier (SiPM) for Time Of Flight Positron Emission Tomography (TOF-PET). IEEE Nucl Sci Symp Conf Rec. 2009.
84. Del Guerra A, et al. Silicon Photomultipliers (SiPM) as novel photodetectors for PET. Nucl Instrum Methods Phys Res A. 2011;648:S232.
85. Choong WS, et al. High-performance electronics for time-of-flight PET systems. JINST. 2013;8:T01006.
86. Roloa MD, et al. TOFPET ASIC for PET applications. JINST. 2013;8:C02050.
87. Streun M, et al. Coincidence detection by digital processing of free-running sampled pulses. Nucl Instrum Methods Phys Res A. 2002;487:530.
88. Siegel S, et al. Simple charge division readouts for imaging scintillator arrays using a multichannel PMT. IEEE Trans Nucl Sci. 1996;43:1634–996.
89. Popov V, et al. Readout electronics for multianode photomultiplier tubes with pad matrix anode layout. IEEE Nucl Sci Symp Conf Rec. 2003.
90. Boisson F, et al. Description and properties of a resistive network applied to emission tomography detector readouts. Nucl Instrum Methods Phys Res A. 2017;872:100–6.
91. Vandenberghe S, Marsden PK. PET-MRI: a review of challenges and solutions in the development of integrated multimodality imaging. Phys Med Biol. 2015;60:R115.
92. Shao Y, et al. Simultaneous PET and MR imaging. Phys Med Biol. 1997;42:1965.
93. Shao Y, et al. Development of a PET detector system compatible with MRI/NMR systems. IEEE Trans Nucl Sci. 1997;44.
94. Pichler B, et al. Studies with a prototype high resolution PET scanner based on LSO-APD modules. IEEE Trans Nucl Sci. 1998;45:1298.
95. Ziegler SI, et al. A prototype high-resolution animal positron tomograph with avalanche photodiode arrays and LSO crystals. Eur J Nucl Med Mol Imaging. 2001;28:136.
96. Grazioso R, et al. APD-based PET detector for simultaneous PET/MR imaging. Nucl Instrum Methods Phys Res A. 2006;569:301.
97. Delso G, et al. Performance measurements of the siemens mMR integrated wholebody PET/MR scanner. J Nucl Med. 2011;52:1914.
98. Maramraju SH, et al. Small animal simultaneous PET/MRI: initial experiences in a 9.4 T microMRI. Phys Med Biol. 2011;56:2459.
99. Peng H, et al. Investigation of a clinical PET detector module design that employs large-area avalanche photodetectors. Phys Med Biol. 2011;56:3603.

100. Kolb A, et al. Technical performance evaluation of a human brain PET/MRI system. Eur Radiol. 2012;22:1776.
101. Judenhofer MS, et al. Simultaneous PET-MRI: a new approach for functional and morphological imaging. Nat Med. 2006;14:459.
102. Catana C, et al. Simultaneous in vivo positron emission tomography and magnetic resonance imaging. PNAS. 2008;105:3705.
103. Kwon SI, et al. Development of small-animal PET prototype using silicon photomultiplier (SiPM): initial results of phantom and animal imaging studies. J Nucl Med. 2011;52:572.
104. Yoon HS, et al. Initial results of simultaneous PET/MRI experiments with an MRI-compatible silicon photomultiplier PET scanner. J Nucl Med. 2012;53:608.
105. Ko, et al. New high performance SiPM PET insert to 9.4-T MR scanner for simultaneous PET/MRI studies. J Nucl Med. 2013:54.
106. Ko, et al. Evaluation of a silicon photomultiplier PET insert for simultaneous PET and MR imaging. Med Phys. 2016;43:72–83.
107. Ko, et al. Simultaneous multiparametric PET/MRI with silicon photomultiplier PET and ultra-high-field MRI for small-animal imaging. J Nucl Med. 2016;57:1309–15.
108. Yamamoto S, et al. Development of a Si-PM-based high-resolution PET system for small animals. Phys Med Biol. 2010;55:5817–31.
109. Yamamoto S, et al. Interference between PET and MRI sub-systems in a siliconphotomultiplier-based PET/MRI system. Phys Med Biol. 2011;56:4147.
110. Yamamoto S, et al. Simultaneous imaging using Si-PM-based PET and MRI for development of an integrated PET/MRI system. Phys Med Biol. 2012;57:N1.
111. Weissler B, et al. MR compatibility aspects of a silicon photomultiplier-based PET/RF insert with integrated digitisation. Phys Med Biol. 2014;59:5119.
112. Wehner J, et al. PET/MRI insert using digital SiPMs: investigation of MR-compatibility. Nucl Instrum Methods Phys Res A. 2014;734:116–21.
113. Wehner J, et al. MR-compatibility assessment of the first preclinical PET-MRI insert equipped with digital silicon photomultipliers. Phys Med Biol. 2015;60:2231–55.
114. Schug D, et al. Initial PET performance evaluation of a preclinical insert for PET/MRI with digital SiPM technology. Phys Med Biol. 2016;61:2851–78.
115. Parl C, et al. A novel optically transparent RF shielding for fully integrated PET/MRI systems. Phys Med Biol. 2017;62:7357–78.
116. Jaszczak RJ. The early years of single photon emission computed tomography (SPECT): an anthology of selected reminiscences. Phys Med Biol. 2006;51:R99–R115.
117. Sajedia S, et al. Design and development of a high resolution animal SPECT scanner dedicated for rat and mouse imaging. Nucl Instrum Methods Phys Res A. 2014;741:169–76.
118. Furenlid LR, et al. FastSPECT II: A Second-Generation High-Resolution Dynamic SPECT Imager. IEEE Trans Nucl Sci. 2004;51:631–5.
119. Walrand S, et al. Evaluation of novel whole-body high-resolution rodent SPECT (Linoview) based on direct acquisition of linogram projections. J Nucl Med. 2005;46:1872–80.
120. Meikle SR, et al. A prototype coded aperture detector for small animal SPECT. IEEE Trans Nucl Sci. 2002;49:2167–71.
121. McElroy DP, et al. Performance evaluation of A-SPECT: a high resolution desktop pinhole SPECT system for imaging small animals. IEEE Trans Nucl Sci. 2002;49:2139–47.
122. Kim H, et al. SemiSPECT: A small-animal single-photon emission computed tomography (SPECT) imager based on eight cadmium zinc telluride (CZT) detector arrays. Med Phys. 2006;33:465–74.
123. Goertzen AL, et al. First results from the high-resolution mouseSPECT annular scintillation camera. Trans Med Imaging. 2005;24:863–7.
124. Barrett HH, Hunter WCJ. Detectors for small-animal SPECT I. Overview of technologies. In: Kupinski MA, Barrett HH, editors. Small animal SPECT imaging. New York: Springer; 2005.
125. Wirrwar A, et al. The optimal crystal geometry for small-field-of-view gamma cameras: arrays or disks? IEEE Nucl Sci Symp Conf Rec. 2000;3:91–3.

126. Wirrwar A, et al. Influence of crystal geometry and wall reflectivity on scintillation photon yield and energy resolution. IEEE Nucl Sci Symp Conf Rec. 1999;3:1443–5.
127. Eisen Y, et al. CdTe and CdZnTe X-ray and gamma-ray detectors for imaging systems. IEEE Trans Nucl Sci. 2004;51:1191.
128. Verger L, et al. "Performance and perspectives of a CdZnTe-based gamma camera for medical imaging," Nuclear Science. IEEE Trans Nucl Sci. 2004;51:3111–7.
129. Meng LJ, et al. Preliminary evaluation of a novel energy-resolved photon-counting gamma ray detector. Nucl Instrum Methods Phys Res A. 2009;604:548–54.
130. Ben-Haim S, et al. Simultaneous dual-radionuclide myocardial perfusion imaging with a solid-state dedicated cardiac camera. Eur J Nucl Med Mol Imaging. 2010;37:1710–21.
131. Montemont G, et al. Evaluation of a CZT gamma-ray detection module concept for SPECT, IEEE Nucl Sci Symp Conf Rec. 2012;4091–4097.
132. Del Guerra A, et al. An integrated PET-SPECT small animal imager: preliminary results. IEEE Trans Nucl Sci. 2000;47:4.
133. Loudos GK, et al. A 3D high-resolution gamma camera for radiopharmaceutical studies with small animals. Appl Radiat Isot. 2003;58:501–8.
134. R. Pani, et al. A novel parallel hole collimator for high resolution SPET imaging with a compact LaBr3 gamma camera. IEEE Nucl Sci Symp Conf Rec. 2008;3824–8.
135. Schramm NU, et al. High-resolution SPECT using multi-pinhole collimation. IEEE Trans Nucl Sci. 2003;50:315–20.
136. Beekman FJ, Vastenhouw B. Design and Simulation of a high-resolution stationary SPECT system for small animals. Phys Med Biol. 2004;49:4579–92.
137. Cao Z, et al. Optimal number of pinhole in multi-pinhole SPECT for mouse brain imaging-a simulation study. Phys Med Biol. 2005;50:4609–24.
138. Funk T, et al. A novel approach to multi-pinhole SPECT for myocardial perfusion imaging. J Nucl Med. 2006;47:595–602.
139. Vanhove C, et al. Three-pinhole collimator to improve axial spatial resolution and sensitivity in pinhole SPECT. Eur J Nucl Med Mol Imaging. 2008;35:407–15.
140. Gullberg GT, et al. Estimation of geometrical parameters for cone beam tomography. Med Phys. 1989;17:264–72.
141. Wang H, et al. Determination of collimator and acquisition parameters for astigmatic SPECT imaging. IEEE Nucl Sci Symp Conf Rec. 1995;2:1116–20.
142. Lewis DP and Tsui BMW Estimation of half fan-beam collimator parameters for brain spect with an l-shaped dual camera spect system. IEEE Nucl Sci Symp Conf Rec. 1998;1052–6.
143. Qi JY, Huesman RH. Effect of errors in the system matrix on maximum a posteriori image reconstruction. Phys Med Biol. 2005;50:3297–312.
144. Mu Z, et al. Recent progress on SPECT imaging with near-field coded aperture collimation: a small animal study. IEEE Nucl Sci Symp Conf Rec. 2010;3450–3.
145. Zeng GL, Gagnon D. Image reconstruction algorithm for a SPECT system with a convergent rotating slat collimator. IEEE Trans Nucl Sci. 2004;51:1.
146. Metzler SD, et al. On-axis sensitivity and resolution of a slit-slat collimator. J Nucl Med. 2006;47:1884–90.
147. Novak JR, et al. Verification of the sensitivity and resolution dependence on the incidence angle for slit-slat collimation. Phys Med Biol. 2008;53:953–66.
148. Song TY, et al. Performance amelioration for small animal SPECT using optimized pinhole collimator and image correction technique. IEEE Nucl Sci Symp Conf Rec. 2005;52:1396–400.
149. Rajaee A, et al. Simulation study of the influence of collimator material on image quality improvement for high energy photons in nuclear medicine using MCNP code. *J Theor Appl Phys*. 2011;4:13–8.
150. Keyes W. The fan beam camera. Phys Med Biol. 1975;20:489–91.
151. Tosswill C. Computerized rotating laminar collimation imaging system US Patent Application 646 (granted December 1977), p. 917–67.
152. Entine G. Cadmium telluride gamma camera. IEEE Trans Nucl Sci. 1979;26:552–8.

153. Gagnon D, et al. Design considerations for a new solid-state gamma-camera: SOLSTICE Proc. IEEE Nucl Sci Symp Conf Rec. 2001;2:1156–60.
154. Van Holen R, et al. Comparing planar image quality of rotating slat and parallel hole collimation: influence of system modeling. Phys Med Biol. 2008;53:19892002.
155. Webb S, et al. Geometric efficiency of a rotating slit collimator for improved planar gamma camera imaging. Phys Med Biol. 1993;38:627–38.
156. Lodge MA, et al. A prototype rotating slat collimator for single photon emission computed tomography. Trans Med Imaging. 1996;15:500–11.
157. Boisson F, et al. Characterization of a rotating slat collimator system dedicated to small animal imaging. Phys Med Biol. 2011;56:1471–85.
158. Boisson F, et al. Assessment of a fast generated analytical matrix for rotating slat collimation iterative reconstruction: a possible method to optimize the collimation profile. Phys Med Biol. 2015;60:2403–19.
159. Jan S, et al. GATE: a simulation toolkit for PET and SPECT. Phys Med Biol. 2004;49:4543–61.
160. Boisson F, et al. Determination of optimal collimation parameters for a rotating slat collimator system: a system matrix method using ML-EM. Phys Med Biol. 2016;61:2302–18.
161. Van der Have F, et al. U-SPECT-II: An ultra-high-resolution device for molecular small-animal imaging. J Nucl Med. 2009;50:599–605.
162. Fresneau N, et al. Design of a serotonin 4 receptor radiotracer with decreased lipophilicity for single photon emission computed tomography. Eur J Med Chem. 2015;94:386–96. https://doi.org/10.1016/j.ejmech.2015.03.017.
163. Van Audenhaege K, et al. Collimator design for a multipinhole brain SPECT insert for MRI. Med Phys. 2015;42:6679.
164. Meier D, et al. A SPECT camera for combined MRI and SPECT for small animals. Nucl Instrum Methods Phys Res A. 2011;652:731–4.
165. Cai L, et al. MRC-SPECT: A sub-500 mm resolution MR-compatible SPECT system for simultaneous dual-modality study of small animals. Nucl Instrum Methods Phys Res A. 2014;734:147–51.
166. Busca P, et al. Simulation of the expected performance of INSERT: A new multimodality SPECT/MRI system for preclinical and clinical imaging. Nucl Instrum Methods Phys Res A. 2014;734:141–6.

# Influence of Animal Handling and Housing on Multimodality Imaging

# 3

David Stout

## 3.1 Introduction: History and Hardware

Medical imaging began first in humans, most famously with Roentgen's X-ray images of his wife's hand and wedding ring [1]. Since then, many types of imaging modalities have been discovered and developed, usually first in humans or cells. The application of imaging methods to preclinical animal research often came later, in part because the immediate application of the technology under development was aimed at improving human health [2, 3]. The more challenging reason for the lag in preclinical use has been that animals used in research are almost always smaller than humans; thus the resolution and sensitivity of these imaging systems need to be correspondingly greater. For example, the difference between the size of a human and a mouse is ~4000-fold, so to study the same concentration of an imaging agent in mice as used in humans requires three orders of magnitude improvement for preclinical versus clinical imaging. There is a trade-off between resolution and sensitivity, and it becomes expensive and difficult to improve both. The manufacturing tolerances required for accurate and precise high-resolution measurements and the signal capturing capability required for high sensitivity and low noise are challenging. Data processing requirements have historically been limiting as well, where the file sizes challenged computer memory and processing capabilities. Improvements in technology and manufacturing have led to the development of imaging systems dedicated specifically to preclinical work in mice and rats [4, 5]. There are now many different commercially available imaging systems with excellent performance and fast data processing capabilities to provide image data quickly and accurately. With proper quality control [6], the equipment is usually very reliable and reproducible.

After initial development of separate PET, SPECT, CT, MR, and optical bioluminescence and fluorescence systems, these modalities began to be combined into

D. Stout (✉)
D&D Design, Culver City, CA, USA



C. Kuntner-Hannes, Y. Haemisch (eds.), *Image Fusion in Preclinical Applications*, https://doi.org/10.1007/978-3-030-02973-9_3

single gantry devices [7]. This was driven in part by the need for anatomical information to help understand the location of metabolic signals and to improve the accuracy of molecular imaging data (scatter and attenuation corrections). When clinical PET/CT systems were developed, they quickly became the standard within just a few years, and now only dual-modality PET/CT and SPECT/CT clinical systems are commercially available. It took longer for this trend to find its way into the preclinical field. Similar convergence has occurred with optical and MR or CT systems [8]. Clinical PET/MR has been well established [9]; however, preclinical multimodality MR systems are only slowly coming to market, with either inserts for PET or SPECT devices into MR magnets [10] or using interchangeable imaging chambers that dock to other separate modality gantries [11].

Initially it was possible to conduct multimodality imaging using a removable imaging chamber to hold the animal while being imaged in separate imaging devices [12]. This design was sometimes prone to movement and positioning errors and required that one chamber work in two or more potentially incompatible gantries from different companies. Consolidation of preclinical hardware companies has led to more compatible gantry designs, since several vendors now offer a wide range of imaging devices and a common animal handling platform has become essential.

Ever since animals were imaged multiple times and in different systems, there has been frequent need for software image co-registration and image fusion. This task is made more complicated because different systems have different hardware and software designs, resulting in variable factors such as fields of view, resolution, and voxel sizes. Often there are software setting choices when configuring the image acquisition and processing that determine the final image parameters. When using separate systems for a combined study, careful selection of the image protocol and using the same setting each time becomes essential. With the development of multimodality systems, the co-registration and fusion process has been taken into account and is often easier or automated. These devices usually provide reproducible positioning and image fusion through hardware co-registration of an animal located inside a chamber or fixture of some kind, usually with a computer-controlled bed movement that enables accurate positioning.

Image fusion can be accomplished using known positioning of the animals or through software co-registration, where known features or signals are aligned. Multimodality gantries make hardware-based fusion easy, provided the imaging parameters are optimized. Using multiple separate gantries, a chamber can still be used in a reproducible manner to make fusion trivial [12]. Where things get tricky is if there is a need to co-register data sets that might have animals in different orientations or attempting to line up multiple data sets to use a common region or analysis tool. Hardware-based options do not require any knowledge of the image content; however, software-based methods require the image data and some method to optimize the match between data sets. This can be done using fiducial markers, though the markers become a distraction in the final image and may require being cropped out. Markers also have to provide a signal in all the modalities being used, which is a challenge with nuclear medicine methods due to half-life loss of signal over time. Another challenge is that functional metabolic data can be quite different than anatomical information; thus it can be quite challenging to match up PET, SPECT, or optical data with MR or CT data. Many groups have worked on this challenging process, with varying degrees of success [13]. There are also several mouse and rat atlases that have been developed with varying degrees of warping and deformation possible to help match up the data sets [14, 15]. Atlases have the advantage of being able to create a uniform orientation of the animals and have predefined regions for determining imaging signal properties. While useful, atlas data are often only suitable for use with normal or near normal anatomy and physiology.

The key to making image fusion work rests with a good reproducible positioning system, thus a well-designed imaging chamber. In addition to positioning, the chamber also helps with reproducible physiological support. Physics, electronics, and the imaging systems are all known, inanimate objects that behave in predictable ways; however what gets imaged within these systems is another story. Animals are not always predicable or behave in the same way each time, despite our best efforts to minimize experimental variables. Some parts of an experiment may have obvious factors that can alter imaging results, such as animal temperature, anesthesia, positioning, injection method, and any chemical or surgical intervention [16]. Other factors are much more subtle and easily overlooked, such as housing conditions [17], diet, circadian effects, cage conditions, and changing frequency. These factors can be divided into two types, acute conditions associated with the imaging experiment that take place within a few hours and chronic conditions that exist in the housing location.

## 3.2 Imaging Chambers: Early Development

The idea of needing a chamber to hold animals only became important when imaging systems were developed for relatively small animals, such as mice and rats. Previously, imaging of larger animals was done by aligning the animal by eye, perhaps using laser alignment lights used in patient scanners. With small-animal systems, there were often few if any tools or options for animal placement. From the beginning of their development, microPET systems had a motorized computer-controlled bed that would move animals in and out of the field of view along the axial direction. The bed was permanently centered in the left/right orientation. Height control motion was possible using a screw or motor that enabled different vertical positions for accommodating rats or mice. As seen in Fig. 3.1, positioning of the mouse or rat in early generation PET scanners was by eye, with fixation using paper tape on cardboard, and there was no temperature control (Fig. 3.1). Image fusion or co-registration mainly depended on either good manual placement or software registration and perhaps image warping.

The metabolic information in PET scans was sometimes difficult to interpret, especially with very specific PET signals where little if any anatomical information was present. It became immediately obvious that there was a need to pair up PET with an anatomical imaging modality, and the most readily available and simplest to integrate was X-ray computed tomography, CT. Fortunately, shortly after

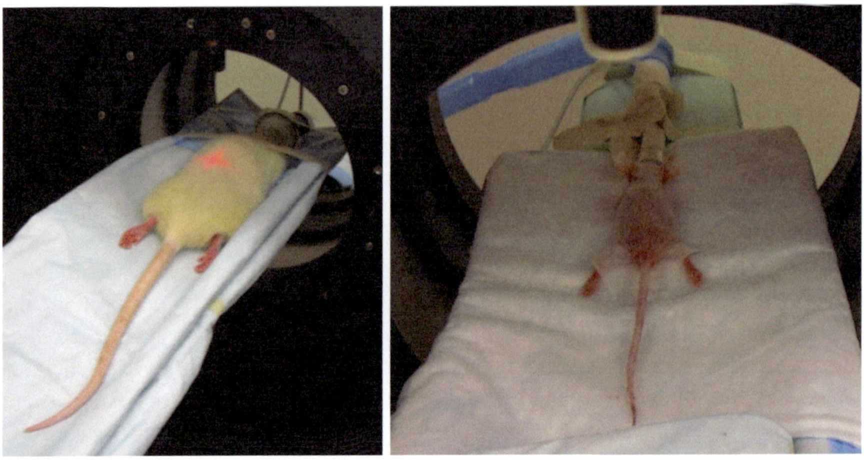

**Fig. 3.1** PET imaging of rats and mice prior to development of imaging chambers. Positioning was done by eye and laser light, with no temperature control or ability to transfer animals to other imaging systems

the development of microPET, small-animal CT systems became available. Since microCT and microPET imaging systems were developed independently, they were initially separate devices, which created a need for a common platform to hold the animals that could facilitate image co-registration. Figure 3.2 shows a chamber designed specifically for mice and rats for use in small-animal PET and CT systems, which provided heating, positioning, gas anesthesia, and pathogen control [12].

One might be tempted to think that the rodent chamber development was aimed at imaging animals in two imaging systems, creating multimodality capability, but that was actually a secondary benefit. The initial reason chambers were developed was to create a controlled gas anesthesia environment and pathogen barrier for the immune-compromised mice that were quickly becoming the most commonly imaged animals, due to their useful ability to grow human tumor cell lines used in oncology research. These animals needed to be protected from the environment and us, since they were missing part or nearly all of their immune system. Preserving the health of the animals through pathogen control, stable anesthesia, and reproducible positioning is essential to maintain the validity of the animal disease model and relevancy of the results to translate into clinical value and also helps with data analysis and image presentation.

Chamber development started with the goals set forth in Table 3.1 and resulted in the chamber shown in Fig. 3.2. This chamber enabled the animal to be positioned within about 1 mm reproducibly, with temperature control provided by a resistive wire heating element [12]. The 30 μm heating wire was so small that there were no artifacts in the CT image. By docking the chamber to a mounting plate fixed to the imaging system gantry platform, the chamber could be reproducibly positioned within the limits of the imaging system resolution (~100 μm for CT).

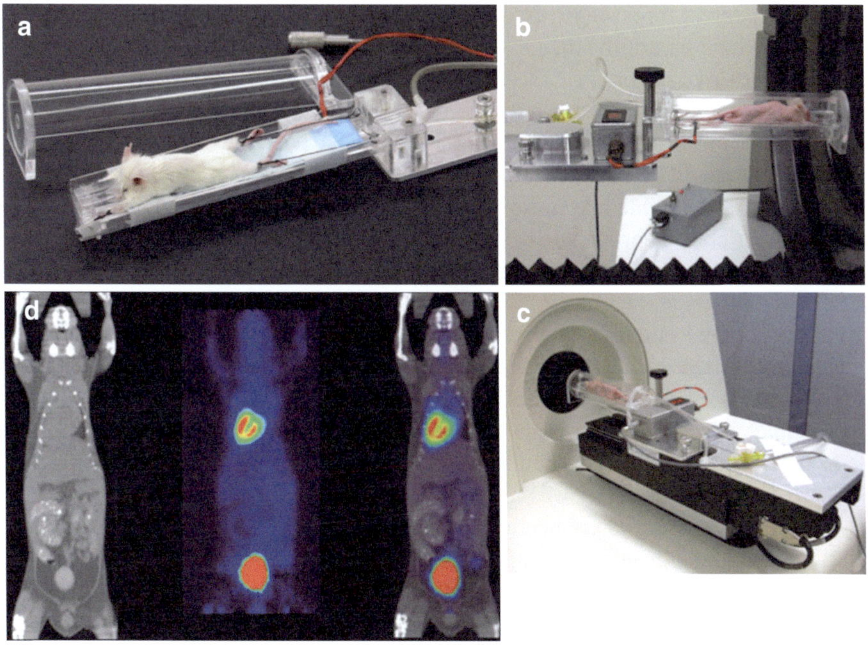


**Fig. 3.2** First-generation imaging chamber (**a**), designed to fit in small-animal PET (**b**) and CT (**c**) gantries. Using a fixed known bed position enabled a simple software co-registration alignment to create fused PET/CT images (**d**)

**Table 3.1** Design elements required for multimodality imaging chambers

| |
|---|
| • Pathogen barrier (isolation chamber) |
| • Gas anesthesia support |
| • Temperature monitoring and control |
| • Materials compatible with multiple modalities |
| • Fixed reproducible positioning in multiple gantries |
| • Reproducible animal positioning for longitudinal studies |
| • Dynamic injection capability |
| • Allow for animal monitoring (visual and sensors) |
| • Easy to use |

The combination of chamber and animal reproducibility coupled with positioning the chamber in the exact same location for each scan enabled a fixed co-registration alignment to be applied to co-register the image data sets (Fig. 3.2d). This was also made possible by using very specific voxel and image matrix size dimensions for PET ($0.4 \times 0.4 \times 0.8$ mm$^3$ voxel size, $128 \times 128$ matrix size) and CT ($0.2 \times 0.2 \times 0.4$ mm$^3$ voxel size, $256 \times 256$ matrix size). These image acquisition and reconstruction parameters meant that the PET and CT images could be overlaid with matching fields of view without the need for resolution matching and software interpolation. Only a fixed offset value was needed to consistently match the image volumes, which was hardware dependent and easy to measure and did not rely on the image content for co-registration. The creation of 3D co-registered fused images became a trivial part of the image assembly, which was straightforward to automate for every set of images. Prior to this hardware/software co-registration, separate image data sets had to be co-registered using software techniques that might be time-consuming and labor intensive. It may also have been necessary to use fiducial markers that could be used as guides to register the data sets. Fiducial markers are not always reproducible in terms of their position if placed on animals. Marker ability to give a consistent signal is challenging due to half-life decay, and they may need to be removed from images for analysis and display.

The chambers shown in Fig. 3.2 were tested and refined over many years and 100,000+ animal studies. They proved to be extremely reliable, providing automated co-registered PET-CT images, ready for analysis without requiring any software co-registration. The fused data sets were automatically scaled to give a reasonable contrast level for both the CT and PET images. There were some problems however, where animals could move while being transported between imaging systems and if the chamber was not properly secured to the mounting plate or the electrical and anesthesia connections were not manually connected. These shortcomings led to the development of the next generation of chambers.

## 3.3 Imaging Chambers: Second-Generation Development

The chamber design and development discussed above was all done by eye and hand, using skilled machinists. This approach was very time-consuming and expensive, often taking 6 months or more to get a new design completed. With the advent of computer-aided design tools (CAD), the next generation of chambers was created using computers. The goals remained the same as before in Table 3.1; however certain improvements were needed. The original chamber required that the heating wire and anesthesia supply and exhaust lines be connected by hand. Sometimes this was overlooked, leading to the animals waking up and the data lost or unnecessary exposure to personnel to the anesthetic gas. The chambers also required manual connection to the mounting plate on the scanner bed. If not mounted correctly, the chamber could be tilted at an angle, requiring manual image co-registration. There was also a problem with animals slipping out of position, most frequently with the nose dropping out of the anesthesia delivery cone. Having the head in two different positions for the two imaging modalities was not something that could be corrected by realignment of the images. A better solution was needed, and it turns out that experience was the best guide with respect to ideal animal placement.

The second generation of imaging chambers uses plug-and-play connection technology (Fig. 3.3). A series of chamber designs were tested, resulting in the final version on the bottom of the left side image. The chamber can be slid into a docking cradle, with the heating and anesthesia connections made automatically. This ensures better positioning and consistent connections, in part since there is both a feel of the chamber clicking into place and a light turns on when the chamber is docked correctly. The chamber is actually just one part of a complete system, which

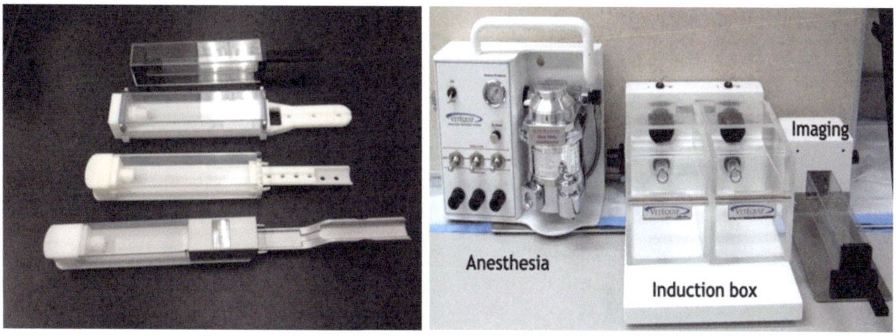


**Fig. 3.3** Second-generation (2009–2015) imaging chamber design evolution (left). Right image shows prototype anesthesia vaporizer, induction boxes, and an early version of a docking station and imaging chamber

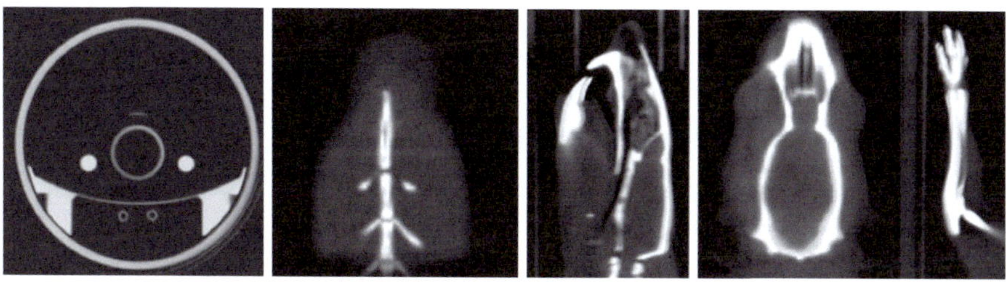

**Fig. 3.4** Second-generation chamber reproducibility testing. Each panel has five overlaid CT images, where the chamber and rat were removed and replaced between each scan. Chamber reproducibility (left image) was <0.21 mm using 0.5 mm voxels. Rat body part positioning, left to right, of sternum, sagittal head view, coronal skull view, and left arm were within 0.28 mm

also includes the anesthesia induction box, docking station for working with the animals in the chamber within a biosafety cabinet and the gantry dock located inside the imaging system. The right side of Fig. 3.3 shows an early design for an anesthesia vaporizer, induction boxes, and docking station with a prototype chamber.

The other improvement to animal positioning and image co-registration was made by removing the nose cone and using a ∩-shaped cowling. The animal can be laid down on the chamber with its head in a natural position. Reproducible trans-axial positioning is provided by the curved bed surface, and axial positioning is provided by where the animal forehead touches the cowling (Fig. 3.4). This design turned out to have much better reproducible positioning than the earlier design and far fewer movement problems when moving between imaging systems and imaging positions within a single gantry. An examination of 100 consecutive studies by multiple different investigators showed an amazingly reproducible positioning capability, without using any tape or restraint devices. All that was needed was the right curvature and place for the nose to touch, plus a little training to have the mouse posture consistently positioned. This improved design resulted in fewer misaligned data sets, improving the image fusion results.

These improvements in chamber technology have been incorporated by nearly every small-animal imaging manufacturer into all in vivo imaging systems. Bruker, Somni, Perkin-Elmer, Sofie, Cubresa, and Mediso now include some kind of animal anesthesia, support, and positioning solution that utilizes some or all of these chamber principles. Chamber designs are matched to the physics and requirements of each imaging modality. For example, a resistive wire heating strip similar to a car window defrost wire is compatible and used for PET, SPECT, and CT systems, whereas this would interfere with MR signals, so warm air or water is used for those systems. Optical imaging often uses a simple box with anesthesia nose ports that is placed in a heated enclosure (e.g., IVIS systems); however other optical systems such as fluorescence tomography (FMT) require the ability to see through the animal, so the chambers have the mouse sandwiched between two panels.

Imaging chambers may also include the ability to monitor animal physiology for monitoring heart rate, respiration, and temperature. Gating for cardiac and respiration allows different parts of these cycles to be separated out for investigation or to remove motion artifacts. The ability to keep animals warm for proper physiology is crucial, as is the need to monitor and maintain an ideal depth of anesthesia. The ability to monitor may be built into the chamber systems from the vendor, such as Bruker, or may be a third-party monitoring systems such as the MouseOx system from Starr Life Sciences.

## 3.4 Imaging Chambers: Future Development

The ability to design parts on a computer coupled with automated machinery to create these parts brought down the cost and time needed to make chambers. The next technology to impact chamber development has been 3D printing, which is ideally suited to preclinical applications. Most 3D printers use plastic-like materials, which are melted and built up layer by layer to create any shape desired. These materials are low density and well suited to PET, CT, SPECT, and MR imaging methods. They can even be used to create custom beds or disposable parts for individual experiments, since the cost to purchase and use these printers is very low and free software to design and print is readily available online. Figure 3.5 shows examples of both a computer-designed and machine-milled imaging chamber assembly (CAD part, left side) and a 3D printed PET/MR bed system for mice. The development and manufacturing time has now shrunk to a matter of days or even hours to draw, print, and use a new design.

Three-dimensional printed parts can be useful for prototyping and customized parts designed for specific studies or even specific animals. For long-term routine use however, these parts can in some cases become brittle and difficult to clean. The infill, layer thickness, and other settings can influence the strength, ability to clean, and density of the chamber parts, all of which may or may not matter with respect to the imaging modality being used. At present, it is not possible to 3D print a clear part, so this method is not yet useful for coverings if it is necessary to see the animals, either for positioning or monitoring purposes.

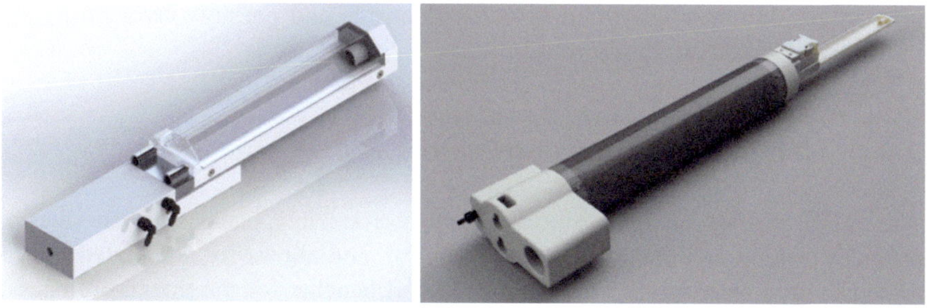

**Fig. 3.5** Left: computer-aided designed (CAD) third-generation PET/CT imaging chamber (left image) made using traditional manufacturing techniques. Right: 3D printed mouse PET/MR chamber. PET/CT chamber uses resistive wire heating, and PET/MR system uses warm air

## 3.5 Image Fusion and Analysis

The use of a chamber enables automated creation of co-registered fusion images that helps tremendously with image analysis. Regions of interest can be determined using any modality and applied to all images. This makes quantitative measurements easier to derive from any image source, whether different modalities or different imaging sessions. One advantage to the reproducible positioning is that often animals are imaged multiple times as part of an experiment, to look at treatment effects or progression of a disease model. With accurate positioning, the images acquired from multiple sessions can be uniformly acquired, enabling the use of predefined regions to be applied to all the image data. The positioning capability of these chambers is far better than the nuclear medicine-based image resolutions, so the co-registration is extremely good. It is also aesthetically desirable to have the animals displayed in the same orientation when creating figures with multiple time point images for publication and presentation.

For image analysis of animals pre- and posttreatment, the regions drawn may need to grow with a changing signal, such as with tumor growth, or they may need to stay the same size when the object is not changing in size over time and treatment. An example of using the same region size would be evaluation of dopaminergic function in a Parkinson's disease model, where the dopamine producing neurons may be treated to reduce the imaging agent uptake to mimic the disease. The reduction of image signal after treatment will appear to be a shrinking of the area, when in reality only a small number of cells will be lost. In this case, the baseline region size and location would be most appropriate to use for a posttreatment evaluation of the signal measurement. Co-registration of the pre and post images thus becomes essential. This can be accomplished or at least greatly aided by using good reproducible positioning hardware, or it may also be necessary to use software for a fixed three-axis alignment ($x$, $y$, $z$), or it may be necessary to use rotational alignment that requires a six-axis alignment or perhaps image warping and interpolation. The more complicated the alignment requirements and processing, the more labor and resource intensive and possibly error prone the process will become.

## 3.6 Handling and Housing Factors

The use of chambers helps position animals reproducibly, enables gas anesthesia, provides heating and perhaps pathogen control, and allows for automated image co-registration and fusion. These are all useful benefits for the imaging process; however, there are other factors that also can significantly alter the physiology and thus metabolic signals being measured using molecular and anatomical imaging techniques. These factors can be broadly divided into two classes, acute factors related to the imaging session and chronic conditions related to the housing conditions.

### 3.6.1 Acute Imaging Factors

From the time animals leave the vivarium until their return after imaging work is completed, the imaging process ideally follows standard operating protocols (SOPs). Nearly all imaging-related work follows the same steps each time, from anesthesia and injections to imaging and recovery. Procedures are followed to ensure consistent conditions and data reproducibility. In the United States, Europe, and many other countries, laws regulating animal research require SOPs that cover all the various procedures and conditions required for any research, including all noninvasive imaging work. These include thermal warming of animals, injection routes and methods, blood and tissue sampling, anesthetics and both biosafety and radiation controls. Monitoring the animal physiology is usually required and in many cases essential to ensure reproducible experimental procedures. Monitoring can be accomplished by visual inspection, remote-sensing devices such as EKG or rectal thermocouple, or remote visual sensing of heart or respiratory rates using a camera.

Temperature control for animals, especially mice with small body mass, is an essential part of keeping the animals healthy and also physiologically stable. Mice quickly adapt the temperature of the surface they are on once anesthetized, so keeping them in a heated induction box, imaging chamber, and recovery area is very important. Mice thermoregulate their body temperature by activating brown adipose tissue (BAT), which can consume up to 50% of all energy used within the body [18]. Preheating of animals for 20–30 min can dramatically reduce the BAT signal commonly seen with energy use imaging agents such as FDG (Fig. 3.6). Mice also control body temperature through controlling blood flow in the tail, which is a common injection site for imaging agents. Warm animals will have more blood flow in the tail, so it will be easier to find the vein, inject, and have the imaging agent delivered to the bloodstream in warm animals with greater blood flow. Because mice rapidly equilibrate to the environment when under anesthesia, the surface the

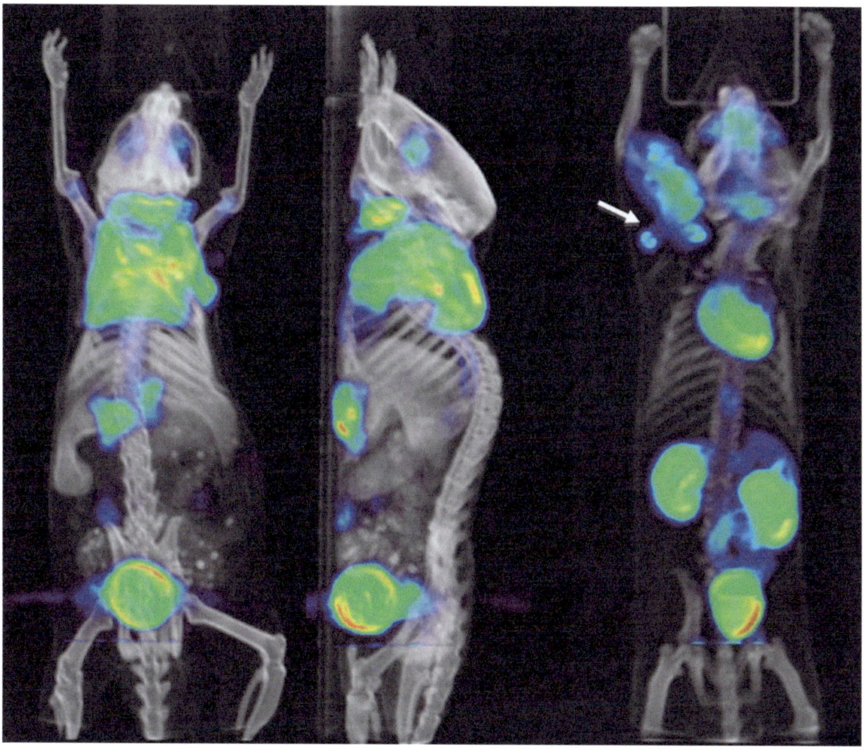

**Fig. 3.6** PET/CT images of FDG in a mouse without warming (left and center images), showing high brown fat (BAT) uptake in the neck and shoulder regions. Right image of a different mouse was acquired with warming, shutting off BAT uptake, allowing visualization of neck tumor, and draining lymph nodes

animal rests upon can be monitored for temperature rather than requiring the use of a more invasive rectal thermocouple. This approach is both safer for the animal and for pathogen control.

There are many other factors that can alter metabolism, including time of day [19], sex of the person handling the animals, fasting, stress, noise, smells, presence of other animals, and other factors beyond the scope of this chapter. There are also factors related to the imaging agent, such as the volume injected, specific activity, pH, presence of alcohol to help dissolve nonpolar molecules, and injection route. These can be trivial to major in terms of their effect, depending on the nature of the experiment. Acclimating the animals and keeping conditions consistent, along with investigating the ideal conditions for your disease model, is an essential part of the experimental design. Reporting these conditions as part of any publication is imperative as well, so that the findings can be replicated [20]. The DICOM committee has expanded the information that can be captured in image header information in an effort to provide a method to record and track this essential information.

The handling conditions, environmental factors, imaging system settings, and housing conditions are all important factors that can influence experimental results. There may be specific parameters necessary to monitor as well, depending on the experimental design. These could include diet, blood sample measurements, metabolite analysis, or behavior. Monitoring may be required over the course of the imaging session or may expend to the length of the entire experiment. For example, longer-lived isotopes may be injected once and followed over days or weeks and may require metabolite analysis to determine the signal information being measured. Good experimental design and protocol adherence becomes key to a successful project.

### 3.6.2 Chronic Imaging-Related Factors

Animals used for experiments spend nearly all their time living in cages within a vivarium, where the veterinary staff typically determines the housing conditions. Vivarium conditions include caging type, room temperature and humidity, lighting, feed, water, cage changing, maximum number of animals per cage, bedding material and amount, and so forth. The choices made can have a huge impact on animal physiology and are known to alter metabolism, histology results, and anatomy [18]. Choices are often made based on cost and the ability to conduct daily heath checks; however, these conditions may make the animal models invalid or variable. Investigators using imaging systems do not normally have control over or pay much attention to the housing conditions, since those are often set by the institution and perhaps thought of as beyond their control or not a source of variability. The variations between vivarium conditions, either within or between institutions, may be one of the biggest causes of irreproducibility of preclinical research results. Even within a single institution, there may be different caging types in use at any one time; thus results may not be replicable simply due to housing location.

When designing an experiment, it is important to consider the housing conditions and their potential impact on the research results. At a minimum, these conditions need to be reported as part of the methods in any publication. In 2015, the National Electrical Manufacturers Association (NEMA) revised the Digital Imaging and Communications in Medicine (DICOM) standards that define all preclinical imaging file formats to expand the header fields to include animal housing and handling conditions, in recognition that these factors play a critical role in experimental results [21].

## 3.7 Summary

The independent development of small-animal imaging systems, coupled with the need to image both metabolism and anatomy while maintaining pathogen control, anesthesia, heating, and reproducibly positioning led to the development of imaging chambers. These chambers enhanced the ability to co-register data sets, use standardized regions of interest, and improved the data analysis process. They enabled hardware-based image co-registration and fusion, which is faster and easier when creating fused images. Succeeding generations of chambers have been integrated into the imaging system design, making it easier to work with animals outside the imaging gantry and simpler to plug the chambers into the gantry with automated connections for anesthesia and heating. The idea of caring for the animal and making physiological conditions suitable and reproducible has now become part of the imaging system design and equipment.

We have also noted how both acute and chronic animal handling and care conditions can drastically alter metabolic imaging data. Animals are used to conduct experiments that would otherwise be unfeasible or impossible in humans, with the implied understanding that the metabolic states of these animals are surrogate representations of human conditions. Animal models of disease states are well known and validated; however, the exact conditions of the validation studies are often not reported in enough detail, nor are they likely to match the conditions in your own facilities. This in part is due to the progressive discovery of new factors that turned out to be important and to the development of new equipment and procedures for housing and care of animals.

Image physics and detailed equipment knowledge have generally been associated with physicists, while animal housing and care procedures are the domain of the veterinary care providers. A researcher using animals to investigate a medical question may not have much shared knowledge with either of these two groups; however, it is the combination of all three of these fields that is essential to acquiring and analyzing meaningful data from living animals. The use of a particular cage type or room temperature decided by the veterinary staff may have a huge unforeseen impact on metabolic data and can lead to irreproducible study results. Likewise, equipment and software changes can alter both qualitative and quantitative imaging results. It is imperative for people working in preclinical research to validate and ensure that the animals are accurately representing an appropriate metabolic state for research work, just as it is to ensure appropriate quality control measurements are consistently acquired to ensure system performance.

## References


1. Glasser O. Wilhelm Röntgen and the early history of the Roentgen rays. San Francisco, CA: Norman Publishing; 1993. p. 25, ISBN: 0930405226.
2. Jaszczak RJ. The early years of single photon emission computed tomography (SPECT): an anthology of selected reminiscences. Phys Med Biol. 2006;51:R99–115. https://doi.org/10.1088/0031-9155/51/13/R07.
3. Portnow LH, Vaillancourt DE, Okun MS. The history of cerebral PET scanning. Neurology. 2013;80:952. https://doi.org/10.1212/WNL.0b013e318285c135.
4. Cherry SR, Gambhir SS. Use of positron emission tomography in animal research. ILAR J. 2001;42(3):219–32. https://doi.org/10.1093/ilar.42.3.219.
5. Zinn KR, Chaudhuri TR, Szafran AA, O'Quinn D, Weaver C, Dugger K, Lamar D, Kesterson RA, Wang X, Frank SJ. Noninvasive bioluminescence imaging in small animals. ILAR J. 2008;49(1):103–15. https://doi.org/10.1093/ilar.49.1.103.

6. Osborne DR, Kuntner C, Berr S, Stout DB. Guidance of efficient small animal imaging quality control. Mol Imaging Biol. 2017;19:485–98. https://doi.org/10.1007/s11307-016-1012-3.
7. Magota K, Kudo N, Kuge Y, Nishijima K, Zhao S, Tamaki N. Performance characterization of the Inveon preclinical small-animal PET/SPECT/CT system for multimodality imaging. Eur J Nucl Med Mol Imaging. 2011;38:742–52. https://doi.org/10.1007/s00259-010-1683-y.
8. Chehade M, Srivastava AK, Bulte JWM. Co-registration of bioluminescence tomography, computed tomography, and magnetic resonance imaging for multimodal *in vivo* stem cell tracking. Tomography. 2016;2:158–65. https://doi.org/10.18383/j.tom.2016.00160.
9. Torigian DA, Zaidi H, Kwee TC, Saboury B, Udupa JK, Cho ZHC, Alavi A. PET/MR imaging: technical aspects and potential clinical applications. Radiology. 2013;267:26–44. https://doi.org/10.1148/radiol.13121038.
10. Judenhofer MS, Wehrl HF, Newport DF, et al. Simultaneous PET-MRI: a new approach for functional and morphological imaging. Nat Med. 2008;14:459–65. https://doi.org/10.1038/nm1700.
11. Sasser TA, Chapman SE, Sanders I, Liepert L, Leevy WM. Cross-platform MRI/PET or MRI/SPECT imaging, and co-registration. 2016. Available at: https://www.bruker.com/fileadmin/user_upload/8-PDF-Docs/PreclinicalImaging/Brochures/T156737_App_Note_Cross-Platform_edit_bur.pdf. Accessed 1 Nov 2016.
12. Suckow CE, Kuntner C, Chow PL, Silverman RW, Chatziioannou AF, Stout DB. Multimodality rodent imaging chambers for use under barrier conditions with gas anesthesia. Mol Imaging Biol. 2009;11:100–6. https://doi.org/10.1007/s11307-008-0165-0.
13. Oliveira FP, Tavares JM. Medical image registration: a review. Comput Methods Biomech Biomed Eng. 2014;17:73–93. https://doi.org/10.1080/10255842.2012.670855.
14. Wang H, Stout D, Chatziioannou A. A deformable atlas of the laboratory mouse. Mol Imaging Biol. 2015;17:18–28. https://doi.org/10.1007/s11307-014-0767-7.
15. Segars WP, Tsui BMW, Frey EC, Johnson GA, Berr SS. Development of a 4D digital mouse phantom for molecular imaging research. Mol Imaging Biol. 2004;6:149–59. https://doi.org/10.1016/j.mibio.2004.03.002.
16. Fueger BJ, Czernin J, Hildebrandt I, Tran C, Halpern BS, Stout DB, Phelps ME, Weber WA. Impact of animal handling on the results of FDG-PET studies in mice. J Nucl Med. 2006;47(6):999–1006.
17. David J, Knowles S, Lamkin D, Stout D. Individually ventilated cages impose cold stress on laboratory mice: a source of systemic experimental variability. J Am Assoc Lab Anim Sci. 2013;52:738–44.
18. David J, Chatziioannou A, Taschereau R, Wang H, Stout D. The hidden cost of housing practices: quantifying the metabolic demands of chronic cold-stress of laboratory mice with non-invasive imaging. Comp Med. 2013;63(5):386–91.
19. Jilge B, Kunz E. The laboratory mouse, Chapter 20. London: Elsevier Academic Press; 2004. https://doi.org/10.1016/B978-012336425-8/50073-X.
20. Stout D, Berr S, LeBlanc A, Kalen J, Osborne D, Price J, Schiffer W, Kuntner C, Wall J. Guidance for methods descriptions used in preclinical imaging papers. Mol Imaging Biol. 2013;12:1–15.
21. NEMA DICOM WG30 Supplemental 187. https://medical.nema.org/medical/dicom/final/sup187_ft_preclinicalanimalacquistitioncontext.pdf.

# Multimodal Optoacoustic Imaging

# 4

Murad Omar, Dominik Soliman, and Vasilis Ntziachristos

## 4.1 Introduction

Imaging is an extremely important tool in modern biomedical research. As the old saying states, "A picture is worth a thousand words." As in many aspects of life, observing events while they occur makes it easier to understand them, thus imaging was and will always be on the forefront of biomedical sciences. By means of imaging, we can diagnose and distinguish between healthy and malignant tissue or make and test hypothesis related to certain biological questions, such as if a certain drug will be able to work as expected or if it doesn't even target the intended organ in the first place.

In the imaging field, four parameters play a major and competing role when selecting the technology of choice: spatial resolution, temporal resolution, sensitivity, and penetration depth (see Fig. 4.1 for a graphical representation). Although imaging techniques have developed tremendously over the last decades, there is still not a single super-technique capable of performing well in all of these parameters. Generally speaking, fully optimizing one of them comes at the cost of the other parameters.

More particularly, either one of these parameters needs to be optimized for a specific top-performance application, or a balance between several parameters needs to be established to cover a broader range of applications. In the last two decades, a technique called optoacoustic (photoacoustic) imaging has been introduced, which is coming close to pushing the boundary of the possible along all

M. Omar · D. Soliman · V. Ntziachristos (✉)
Chair of Biological Imaging, Technische Universität München, Munich, Germany

Institute of Biological and Medical Imaging, Helmholtz Zentrum München, Neuherberg, Germany
e-mail: v.ntziachristos@tum.de



C. Kuntner-Hannes, Y. Haemisch (eds.), *Image Fusion in Preclinical Applications*, https://doi.org/10.1007/978-3-030-02973-9_4

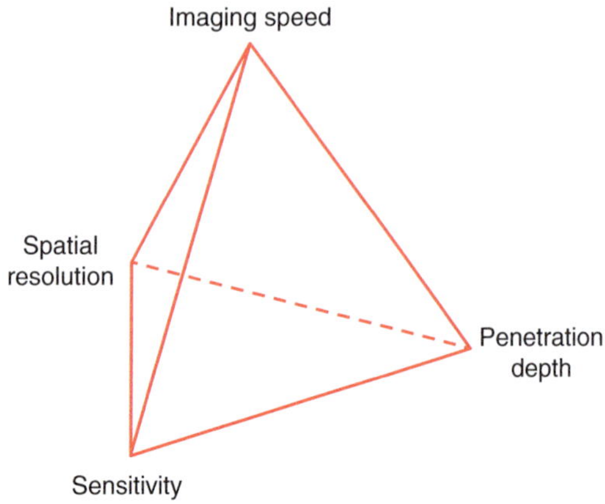


**Fig. 4.1** Imaging pyramid showing the different parameters that can be optimized when designing an imaging system. Generally, if one parameter is maximally optimized, compromises in the other dimensions need to be made

four dimensions in the parameters space. This technique is derived from optical imaging. Thus, we will first discuss the case for optical imaging in the biomedical sciences.

### 4.1.1 Optical Imaging

Optical imaging is a very powerful and widely applied technique. It has been used for several centuries to gain new biological insights, to study and understand disease, and even to diagnose patients. The strength of optical imaging stems from two facts. Firstly, we as human beings tend to better understand what we see, and, thus, looking at an optical image facilitates the relation of the observed to a certain function or disease. Secondly, biological tissue commonly changes its color based on the underlying organ or the disease of interest. Therefore, it is only natural to expect that such a contrast will deliver valuable information about biological processes, such as disease developments, function, and metabolism, and generate images of the bio-distribution of a certain molecular agent.

Optical imaging methods can be divided into two main categories: microscopic and tomographic modalities. In the microscopic case, usually a focused beam of light is used to collect information from the sample of interest. This can be achieved either point by point or line by line, depending on the specific configuration. Such a method, as can be already understood from its name, delivers images with spatial resolutions high enough to observe single cell or even subcellular features. On the other hand, these techniques can only collect information from within a thin layer (around the optical focus) of the examined sample and can thus only image a small volume within a reasonable amount of time. The major limiting factor in the case of penetration depth is the strong scattering of visible light that is experienced in biological tissue [1]. Because of this optical scattering, it becomes impossible to focus a beam of light within a biological sample beyond a few hundred micrometers of depth. Consequently, to use a microscopic method for preclinical experiments, it is necessary to (1) image only the superficial layers in an animal (e.g., the skin); (2) replace the part of the animal between the organ of interest and the microscope with a window chamber, which is widely used in neuroimaging or cancer research; (3) use cell cultures if possible, which can also deliver multiple insights in many processes and models; or (4) clear the organ of interest, which, however, limits the applicability to ex vivo studies.

The other category of optical imaging techniques is so-called tomographic methods. Examples of such techniques include diffuse optical tomography (DOT) and fluorescence molecular tomography (FMT). In the tomographic case, instead of using a focused beam, a broad illumination is employed. In order to obtain a reconstruction of the underlying structure or function of the examined sample, light propagation is modeled using the optical diffusion or the radiative transfer equation [2]. Based on these models, an inverse optimization problem is solved to reconstruct the final images. In other words, mathematical techniques are used for image reconstruction. As the models are based on optical diffusion, the imaging depth in tissue is no longer limited to the first few hundreds of micrometers, rather it becomes possible to image several millimeters or even centimeters of depth in tissue. Additionally, optical tomography allows to image the function of a living tissue by using an excitation at appropriate wavelengths and thus to deduce information about the hemodynamics and metabolism [3]. Finally, it is possible also to inject molecular fluorescence agents, which, instead of merely visualizing structure or function, enable the imaging of the bio-distribution of such an agent, which provides high specificity at centimeters of depth [2]. Nonetheless, the cost to pay for this extended imaging depth is spatial resolution. For example, in the case of FMT, a spatial resolution of 1–2 mm is achieved in the best-case scenario [2].

### 4.1.2 Alexander Graham Bell and the Discovery of the Optoacoustic (Photoacoustic) Effect

Alexander Graham Bell, who is more famous for his invention of the phone, has discovered the optoacoustic (photoacoustic) effect in 1883. The optoacoustic effect is based on the conservation of energy and momentum, where optical energy in particular and electromagnetic energy in general are absorbed by certain molecules that absorb light at specific wavelengths (chromophores) in the specimen. Consequently, this absorption results in an instantaneous raise in the local temperature around the chromophores, leading to a thermoelastic expansion and a generation of acoustic waves. Thus, energy is transformed from electromagnetic waves to heat and finally to sound (see Fig. 4.2 for a graphical illustration of this process).

Bell noticed that acoustic waves are generated when a thin membrane is shined on with modulated light. To capitalize on this newly discovered effect, he suggested building a wireless link between two persons: one is listening to the generated acoustic waves, while the other one is modulating the shining light. This was effectively the first wireless communication link in history, called photophone. Although wireless communication was a brilliant idea, it was not practical as it only worked

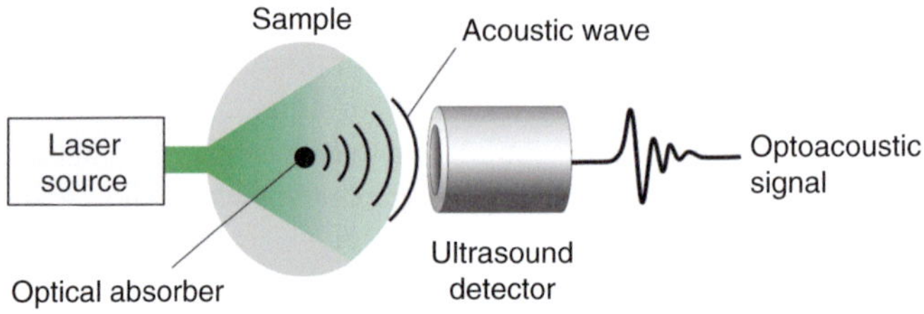


**Fig. 4.2** Optoacoustic imaging principle. The sample is illuminated by a pulsed laser source. Chromophores in the sample absorb optical energy, which is converted into heat via non-radiative processes. This transient temperature rise causes a local pressure build-up, which propagates through the sample as an acoustic wave and which is finally measured by an ultrasound detector (e.g., transducer)

at short distances and needed a line of sight between the sender and receiver. After the invention of the photophone, the optoacoustic effect was exploited in the 1930s to characterize and spectrally measure certain gases. In fact, this has been the major use of the optoacoustic effect until the early 1980s, when Theodore Bowen proposed the applicability of this effect for biological imaging [4]. Finally, optoacoustic imaging started to pick up traction in the 1990s with the introduction of sufficiently strong laser sources and sensitive ultrasonic detectors.

### 4.1.3 The Case for Optoacoustic Imaging: Listening to Light

As we have seen in the previous discussion, optical imaging yields strong and rich contrast, but the spatial resolution starts degrading dramatically beyond the first few hundred micrometers of depth. To overcome this limitation, optoacoustic imaging has been introduced. In optoacoustics, instead of relying on an optical focus to get the desired information from a pixel or voxel, the information about the location is derived from the acoustic propagation time. In such a case, ultrasonic waves, which are 2–3 orders of magnitude less susceptible to scattering than light, are used to generate an image. Because of this usage of ultrasonic waves, it becomes possible to image several centimeters into tissue with an adequate spatial resolution in the range of 100 μm, which is at least an order of magnitude better than what could be achieved with pure optical imaging techniques deep inside the tissue. Moreover, because ultrasonic detectors can be responsive to all kinds of frequency ranges depending on the material and fabrication, it is possible to trade spatial resolution with imaging depth. More specifically, it is possible to partially sacrifice the penetration depth and to get images with a resolution of 20 μm in return, which again is much better than what pure optical techniques can deliver from deeper layers of tissue.

Essentially, optoacoustics delivers high optical contrast side by side with ultrasonic resolution. Consequently, it takes the best features of the two worlds and combines them into a single imaging modality. Additionally, because it facilitates the parallelization of acoustic detection, it is possible to acquire optoacoustic images at video rates or higher. Thus, the optical imaging capabilities have been pushed by the optoacoustic modality along three of the dimensions of the parameter space: penetration depth, spatial resolution, and imaging speed.

In the next sections, we will introduce the basic principles of optoacoustic imaging, where we derive the governing equations and describe the limitations on spatial resolution and imaging depth. Subsequently, we dive into the different scales achievable with optoacoustic imaging; those are the macroscopic, mesoscopic, and microscopic scales. Additionally, we describe the different possibilities for multimodal imaging in the macroscopic, mesoscopic, and microscopic cases based on selected example applications. Finally, we conclude the chapter with a summary and an outlook.

## 4.2 Basic Principles

### 4.2.1 Initial Pressure Generation

The origin of optoacoustic signals is the absorption of laser light by tissue chromophores, followed by the transient heating of a local tissue region that leads to a fractional volume expansion, which can be expressed as [5]

$$\frac{dV}{V} = -\kappa p(\vec{r}) + \beta T(\vec{r}). \tag{4.1}$$

Here, $p(\vec{r})$ denotes the initial pressure change (Pa), $T(\vec{r})$ is the temperature change (K, typically below 0.1 K), $\kappa$ represents the isothermal compressibility (~5 × $10^{-10}$ $Pa^{-1}$), and $\beta$ denotes the isobaric thermal expansion coefficient (~4 × $10^{-4}$ $K^{-1}$) [5]. For efficient optoacoustic pressure generation, the laser pulses have to be sufficiently short. More precisely, the pulse width $\tau$ has to be shorter than both the thermal relaxation time $t_{th} = l_h^2 / \alpha_{th}$ and the stress relaxation time $t_{st} = l_h/c$, which represent the time that it takes the heat and built-up stress, respectively, to propagate out of the heated volume. The length of the heated volume is represented by $l_h$ (m), whereas $\alpha_{th}$ denotes the thermal diffusivity ($m^2/s$), and $c$ is the speed of sound (~1500 m/s in tissue). Usually, $t_{st}$ is much smaller than $t_{th}$ (e.g., for an absorber diameter of 20 μm, $t_{th} \approx 2.6$ ms and $t_{st} \approx 13$ ns).

Under thermal and stress confinement, the fractional volume change can be neglected, and the initial optoacoustic pressure rise can be written as

$$p_0(\vec{r}) = \Gamma \eta_h \mu_a(\vec{r}) \phi(\vec{r}), \tag{4.2}$$

where $\Gamma$ represents the dimensionless and tissue-dependent Grüneisen parameter $\Gamma = \beta/\kappa\rho C_V$, $\rho$ is the mass density ($kg/m^3$), $C_V$ denotes the specific heat capacity (J/(kg K)), $\eta_h$ is the heat conversion efficiency, $\mu_a(\vec{r})$ denotes the optical absorption coefficient (1/m), and $\phi(\vec{r})$ represents the optical fluence ($J/cm^2$). Because optoacoustic signal generation originates from non-radiative electronic relaxation, $\eta_h$ and thus $p_0(\vec{r})$ are increased for chromophores with a low fluorescence quantum yield (i.e., the ratio of the number of emitted to absorbed photons).

### 4.2.2 Optoacoustic Wave Equation and Forward Solution

The propagation of the generated pressure as a bipolar acoustic wave under the condition of thermal confinement and in an acoustically homogeneous medium is described by the optoacoustic wave equation [5–7]:

$$\left(\nabla^2 - \frac{1}{c^2}\frac{\partial^2}{\partial t^2}\right)p(\vec{r},t) = -\frac{\Gamma}{c^2}\frac{\partial}{\partial t}H(\vec{r},t). \tag{4.3}$$

Here, $H(\vec{r},t)$ denotes the heating function (J/(cm$^3$ s)), which is defined as the thermal energy deposited in the tissue through optical absorption per unit time and unit volume [5]. It is defined as

$$H(\vec{r},t) = \eta_h \mu_a(\vec{r})\Phi(\vec{r},t), \tag{4.4}$$

where $\Phi(\vec{r},t)$ is the optical fluence rate $\Phi(\vec{r},t) = \partial\phi(\vec{r},t)/\partial t$ (J/(cm$^2$ s)), which corresponds to an average value of the optical intensity in a turbid diffusive medium [8]. Due to the time derivative in (4.3), only time-variant heating can result in the generation of optoacoustic signals, i.e., pulsed or intensity-modulated laser sources.

The forward solution $p(\vec{r},t)$ to the optoacoustic wave equation, i.e., the total pressure measured at position $\vec{r}$ and time instant $t$, can be found by using the Green's function approach to be

$$p(\vec{r},t) = \frac{\Gamma}{4\pi c^2}\frac{\partial}{\partial t}\iiint_V \left[\frac{d^3\vec{r}'}{\left|\vec{r}-\vec{r}'\right|}H(\vec{r}',t')\right]_{t'=t-\left|\vec{r}-\vec{r}'\right|/c}. \tag{4.5}$$

The heating function can be decomposed into a position- and time-dependent term, whereas the latter can be approximated as a delta function in the case of sufficiently short laser pulses: $H(\vec{r},t) = H_r(\vec{r})\cdot\delta(t)$. Then, the forward solution simplifies to

$$p_{\delta t}(\vec{r},t) = \frac{\partial}{\partial t}\left[\frac{t}{c}\iint_S d\Omega\, p_0(\vec{r}')\right]_{\left|\vec{r}-\vec{r}'\right|=ct}. \tag{4.6}$$

According to (4.6), the total pressure signal $p_{\delta t}(\vec{r},t)$ is obtained by integrating the contributions of all sources located at spherical shells with radius $\left|\vec{r}-\vec{r}'\right|$ centered at $c \cdot t$ (spherical Radon transform).

In the special case of a spherical absorber with the same material properties as the surrounding medium, an analytical expression of the forward solution can be obtained as

$$p_{\delta t,\mathrm{sph}}(\vec{r},t) = p_0(\vec{r})\Theta\left(\frac{d}{2} - |\Delta r - ct|\right)\frac{\Delta r - ct}{2\Delta r}, \tag{4.7}$$

where $\Delta r = |\vec{r} - \vec{r}_{\mathrm{a}}|$, $\vec{r}_{\mathrm{a}}$ is the position of the absorber center and $d$ is the absorber diameter [6]. $\Theta(x)$ is the Heaviside function, which is defined as

$$\Theta(x) = \begin{cases} 0, & x < 0 \\ 1, & x \geq 0 \end{cases}. \tag{4.8}$$

The corresponding optoacoustic signals yield a characteristic bipolar N-shape, as illustrated for absorber diameters of 10, 20, and 40 μm in Fig. 4.3a. Under stress confinement, the amplitude and the duration of the signals are proportional to the source diameter. For solid absorbers with a higher speed of sound and mass density, the optoacoustic signals deviate from the perfect N-shape, yielding a reduced width and several trailing amplitude oscillations [9, 10].

Finally, the solution for a finite laser pulse duration is calculated by convolving $p_{\delta t}(\vec{r},t)$ with the temporal pulse profile $H_{\mathrm{t}}(t)$:

$$p(\vec{r},t) = \int_{-\infty}^{\infty} d\tau\, H_{\mathrm{t}}(t-\tau)\, p_{\delta t}(\vec{r},\tau). \tag{4.9}$$

If the absorber is sufficiently small compared to the resolution of the used ultrasound detector, it can be approximated as a point source (i.e., $H_{\mathrm{r}}(\vec{r}) = \delta(\vec{r})$), and the forward solution further simplifies to

$$p_{\delta t,\delta r}(r,t) = \frac{\Gamma}{4\pi c^2 r}\frac{\partial}{\partial t}\delta\left(t - \frac{r}{c}\right). \tag{4.10}$$

This is especially relevant for optoacoustic microscopy applications, which will be discussed later.

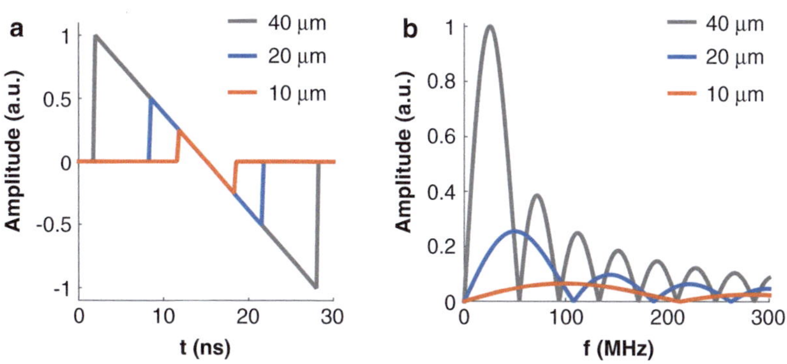


**Fig. 4.3** Simulation of optoacoustic signals from 10, 20, and 40 μm absorbers based on (4.7). (**a**) Time courses of the simulated optoacoustic signals yielding the characteristic N-shape. (**b**) Corresponding frequency spectra

### 4.2.3 Optoacoustic Frequencies

Optoacoustic signals are generally composed of a continuous set of frequencies in the MHz range, the bandwidth of which being inversely proportional to the duration of the time signals and thus to the size of the optical absorber. Figure 4.3b presents the frequency spectra of the simulated N-shaped time signals from 10, 20, and 40 μm absorbers shown in Fig. 4.3a. As can be seen, the smaller absorbers generate broader spectra, whereas all frequency spectra show a main lobe and several side lobes with intermediate minima.

The frequency content of optoacoustic signals is essential for the respective imaging application, as the highest detectable frequencies determine the highest achievable spatial resolution (an exception is the lateral resolution in optoacoustic microscopy, which is governed by the optical focusing capabilities, as will be discussed later). Consequently, the differentiation between macroscopy (tomography), mesoscopy, and microscopy based on the achieved spatial resolutions can be related to the respective ranges of optoacoustic frequencies that are typically detected.

### 4.2.4 Acoustic Attenuation

The detection of different optoacoustic frequency ranges has not only implications on the achievable spatial resolution but also on the maximum imaging depth. More specifically, the lower the highest detectable optoacoustic frequencies, the larger the maximum penetration depth and vice versa. The reason for this relationship is a process known as acoustic attenuation, which refers to a partial energy loss of acoustic waves as they propagate through a medium. This effect originates from frictional losses through the molecules of the medium and from heat diffusion between adjacent volumes of differing temperature [11]. The acoustic attenuation effect is generally stronger for higher than for lower frequencies and can be described by the following power law:

$$p(f,r) = p_0 \, e^{-\alpha_0 |f|^n r}. \tag{4.11}$$

Here, $p(f, r)$ represents the component of the pressure amplitude with frequency $f$ after propagating a distance $r$ in the medium, $p_0$ is the initial pressure amplitude, $\alpha_0$ denotes the attenuation constant in units of *nepers*, and $n$ is a constant that depends on the medium properties [12]. In tissue, the attenuation constant has an average value of $\alpha_{0,\mathrm{t}} = 0.5\ \mathrm{dB/MHz}$ and an exponent of $n_\mathrm{t} \approx 1$. The respective values for water are $\alpha_{0,\mathrm{w}} = 0.000217\ \mathrm{dB/MHz^2}$ and $n_\mathrm{w} \approx 2$ [13].

Consequently, acoustic attenuation in tissue is the limiting factor regarding the highest detectable optoacoustic frequencies and thus the achievable spatial resolution with a particular optoacoustic imaging modality.

### 4.2.5 Tomographic Reconstruction Techniques

The goal of tomographic reconstruction in optoacoustic imaging is to find the spatial distribution of the initial pressure amplitudes $p_0(\vec{r})$ and thus the optical absorption $\mu_a(\vec{r})$ (assuming a constant optical fluence rate, see (4.2)) in two or three dimensions from the pressure waves $p(\vec{r}_i,t)$ measured at different positions $\vec{r}_i$ (so-called inverse problem) [14]. The available reconstruction techniques can be divided into two main categories: analytical and numerical methods.

The most commonly used analytical reconstruction technique is the backprojection method, which is easy to implement, fast, and memory efficient. It is based on the projection of the optoacoustic signals measured at positions $\vec{r}_i$ onto spherical shells (in the 3D case) centered at $\vec{r}_i$ by exploiting the time-of-flight (TOF) information of the time-domain signals and the known or assumed speed of sound $c$ of the medium. Consequently, for each time instant $t_i$, the respective part of the recorded signals is projected back to all positions from which they could have emerged (i.e., spheres with radius $c \cdot t_i$, according to (4.6)), and all contributions are summed up in the reconstruction volume. For the most common detection geometries, i.e., planar, spherical, and cylindrical, the following backprojection formula [15] provides an analytical expression for the initial pressure $p_0$ at position $\vec{r}$:

$$p_0(\vec{r}) = 2\iint_{\Omega} \frac{d\Omega_i}{\Omega}\left[p(\vec{r}_i,t) - t\frac{\partial}{\partial t}p(\vec{r}_i,t)\right]_{t=|\vec{r}-\vec{r}_i|/c}. \tag{4.12}$$

The factor $d\Omega_i/\Omega$ is a solid angle weighting factor corresponding to detection position $i$, which is constant in the far-field approximation (i.e., $|\vec{r}-\vec{r}_i|$ is much larger than the imaged objects), and $\Omega$ is the total solid angle ($2\pi$ for perfect planar and $4\pi$ for closed spherical and cylindrical geometries).

Even though the backprojection method is fast and simple to implement, it does not easily allow for other parameters, such as the detector properties, to be incorporated. On the other hand, numerical (also called model-based) reconstruction techniques facilitate the modeling of detector properties and other physical effects by using a system-dependent forward model and by discretizing the forward problem spatially and temporally. The discretized forward problem can be written as a matrix vector equation of the form

$$\vec{p} = \boldsymbol{M} \cdot \vec{H}, \tag{4.13}$$

where $\vec{p}$ is the vector of measured pressure at different positions and time instants, $\vec{H}$ is the vector containing the values of deposited energy (i.e., optical absorption) in the reconstruction volume, and $\boldsymbol{M}$ represents the model matrix [14]. The goal of all model-based reconstruction algorithms is to invert (4.13) in order to solve for $\vec{H}$, e.g., by iteratively minimizing the following expression to obtain $\vec{H}_{\text{sol}}$:

$$\vec{H}_{\text{sol}} = \arg\min_{\vec{H}} \left\|\vec{p} - \boldsymbol{M} \cdot \vec{H}\right\| + \vec{\delta}_{\text{reg}}. \tag{4.14}$$

Here, ||·|| refers to the $l_2$ norm, and $\vec{\delta}_{\text{reg}}$ is a regularization term that is necessary in the case of ill-posed inverse problems [14]. The disadvantage of model-based reconstruction methods is their high computational demand in terms of time and memory. Therefore, model-based approaches are typically used in optoacoustic tomography applications, where a mere hundreds of measurements are performed per image. On the other hand, high-resolution (especially mesoscopic) optoacoustic modalities that record thousands or even hundreds of thousands of projections per scan usually rely on the backprojection technique for reconstruction.

## 4.3 Optoacoustic Macroscopy (Tomography)

### 4.3.1 Introduction

The word "tomography" originates from the ancient Greek words "tomos" (slice, section) and "graphō" (to write) and refers to imaging by sections or sectioning through the use of any kind of penetrating waves or mechanical method [16]. Hence, optoacoustic tomography refers to optoacoustic sectional imaging of biological organisms. Applications of optoacoustic tomography include the imaging of vascular structures, physiological readings, bio-distributions of targeted optical contrast agents, kidney function, cerebrovascular activity, and tumor hypoxia [17].

### 4.3.2 Workings of Optoacoustic Tomography

Optoacoustic tomography (OAT)[1] is, similar to all optoacoustic methods, based on the optoacoustic effect, where a laser pulse illuminates the whole animal or imaged section at once, followed by a collection of the generated optoacoustic signals, usually by an array of ultrasonic detectors (e.g., transducers). Finally, mathematical methods or models are used to generate a useful image from the recorded signals. To collect signals from whole animals, e.g., mice, low megahertz ultrasonic frequencies, typically 3–10 MHz, are used in tandem with near-infrared illumination to penetrate through 1–5 cm of tissue. Such a deep penetration allows for the noninvasive acquisition of, e.g., cross-sectional slices of mice.

The type of collected information can be divided into three categories: anatomical, functional, and molecular. Anatomical information visualizes, for example, the different organs of a mouse, such as the kidney and the liver, and gives information about biological structures. This anatomical information requires only a single

[1] Although the word "tomography" implies the use of mathematical methods for image reconstruction, in the context of imaging, it also generally refers to whole-body imaging at a macroscopic scale. Hence, we will follow this terminological tradition here and use "macroscopy" and "tomography" synonymously.

illumination wavelength. If multiple wavelengths are available, it is possible to perform functional and molecular imaging based on the distinct absorption coefficients of chromophores at different wavelengths. Functional imaging shows the activity and the metabolism of an organ or tissue and is facilitated through imaging the oxygenation state of tissue rather than its shape. As previously mentioned, this is achieved through imaging the tissue at multiple wavelengths in what is called multispectral imaging. Similarly, molecular imaging is enabled by imaging the distribution of a contrast agent inside the tissue or the bio-distribution of the molecular agent. This kind of visualization is achieved by injecting a molecule that has a defined absorption spectrum and mathematically unmixing it from the images obtained at different wavelengths.

Finally, in order to perform sectional imaging, two approaches can be used. Firstly, it is possible to have the ultrasonic detectors collect signals from all directions, to reconstruct a three-dimensional (3D) image, and to take a cross section from the final reconstruction. Alternatively, it is possible and generally faster to use cylindrically focused arrays, i.e., arrays that collect signals only from a single plane. In such a way, only signals originating from the focal plane are detected, and a tomographic image is generated. If multiple planes are required, the array or the animal can be translated parallel to the axis of the array, and the reconstructed images can be stacked in order to form a 3D volume.

### 4.3.3 Example Setups

Among the multitude of optoacoustic tomography systems available worldwide, the setup developed at the Technical University of Munich (TU Munich) is one of the most suitable and versatile for the imaging of small animals. The system (see Fig. 4.4a) is based on a technology termed multispectral optoacoustic tomography

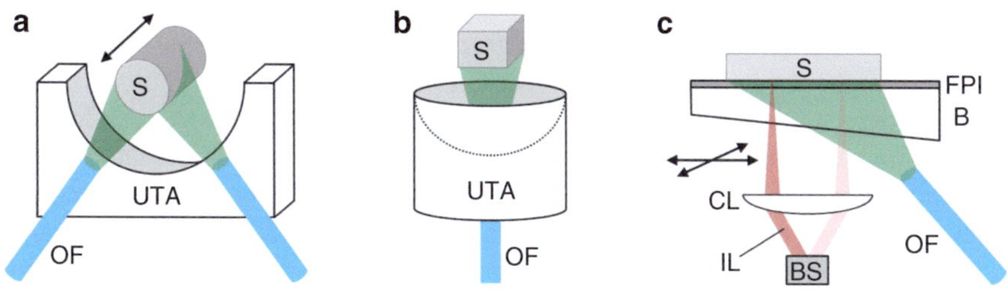


**Fig. 4.4** Optoacoustic macroscopy (tomography) implementations. (**a**) Multispectral optoacoustic tomography (MSOT) based on a transducer ring array and ring illumination [30]. (**b**) Spherical transducer array with a central hole for illumination [18]. (**c**) Optical raster-scanning of an interrogation laser beam within a Fabry-Pérot resonator [29]. Abbreviations: *B* backing stub, *BS x–y* beam scanner, *CL* convex lens, *FPI* Fabry-Pérot interferometer, *IL* interrogation laser, *OF* optical fiber (bundle), *S* sample, *UTA* ultrasound transducer array. *Green color*: laser illumination

(MSOT). The central part is a cylindrically focused transducer ring array, which acquires the signals from a single plane, and a fast tunable near-infrared laser, which can be tuned in the spectral range of 700–1000 nm. The fast tunability of the laser allows the system to image rapid dynamic processes within a living animal, such as metabolism and contrast agent diffusion. In addition to the large number of possible applications, some of which will be discussed in the next subsection, the high wavelength-tuning speed enables the imaging of dynamic processes without the need for co-registration of the acquired images. Other examples of optoacoustic tomographic geometries include spherical transducer arrays [18] and optical raster-scanning of an interrogation laser beam within a Fabry-Pérot resonator [19], as illustrated in Fig. 4.4b, c.

### 4.3.4 Example Applications

Optoacoustic tomography offers a broad range of noninvasive preclinical and biomedical applications. In this section, we will discuss three exemplary applications: neuroimaging, cancer imaging, and the imaging of the bio-distribution of contrast agents. Other applications include the imaging of tumor heterogeneity [20], inflammation [21], and cardiovascular diseases [22, 23].

1. *Neuroimaging*: Because optoacoustic tomography is capable of imaging oxygenation states of tissue, it is ideally suited for the monitoring of activity and metabolism, which are directly linked to oxygenation states inside the organs. An interesting related application is the imaging of neuro-activity inside a mouse brain, which is similar to the blood-oxygen-level dependent (BOLD) imaging technique used in magnetic resonance imaging (MRI) but faster. Hence, it is possible to image not only the resting state of the brain but also faster neuro-activities [24–26]. An example of imaging the perfusion of a molecular agent inside a mouse brain is shown in Fig. 4.5a [18].
2. *Cancer imaging*: Optoacoustic tomography in general and MSOT in particular allow for the imaging of cancer in mouse models. This technology can be used for visualizing cancer heterogeneity [20], cancer hypoxia [27], and the reaction of cancer to different kinds of therapy, such as chemotherapy [28]. The capabilities of the system enable the imaging of whole tumors and not only the superficial regions as commonly done in microscopy. Additionally, because MSOT is capable of noninvasive imaging, there is no need for a window chamber, which is less stressful for the animal. Figure 4.5b shows an example application of optoacoustic macroscopy in the monitoring of tumor development [29].
3. *Imaging of bio-distribution*: Because of its unique capability at distinguishing between the spectra of different molecules, it is possible to image the bio-distribution of a certain molecule with MSOT and to distinguish it from the surrounding tissue (see Fig. 4.5c) [30].

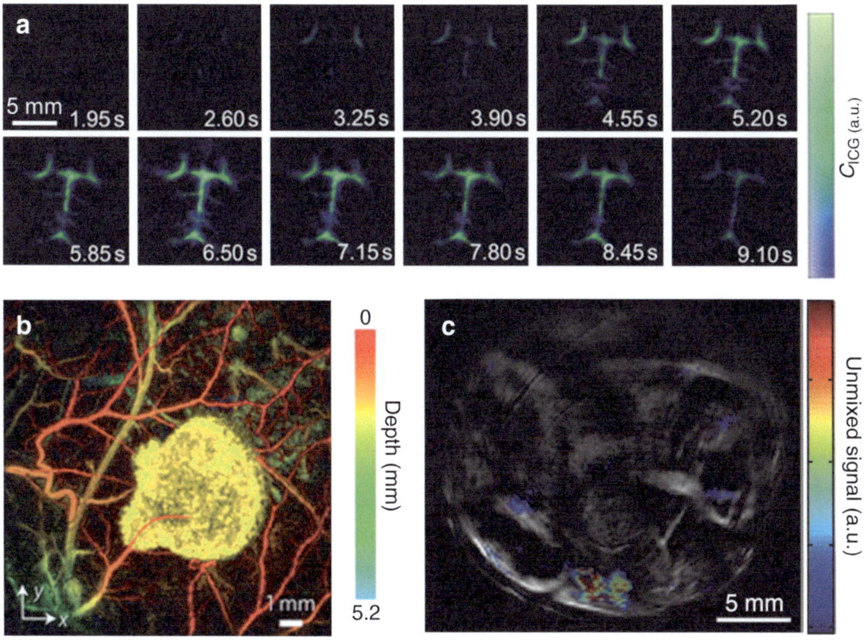


**Fig. 4.5** Examples of optoacoustic macroscopy (tomography) applications. (**a**) Monitoring of the diffusion of a molecular agent (ICG) in a mouse brain over time using a spherical transducer array. Reprinted with permission from [18]. (**b**) Imaging of a solid tumor together with the vascular bed in a mouse model of cancer in vivo using a Fabry-Pérot resonator. Reprinted with permission from [29]. (**c**) Cross section from a mouse body acquired using MSOT, showing the unmixed signal from ICG (*color*) over a single wavelength image (*gray*). Reprinted with permission from [30]

### 4.3.5 Multimodal Tomographic Imaging

To increase the accessible range of contrast and to collect more comprehensive information from the examined samples, optoacoustics offers the possibility for multimodal imaging. Since in optoacoustics the sample is excited by light and ultrasound is detected, the most obvious and simplest multimodal combination would be either to have an optical or an acoustical add-on to the OAT modality. Generally speaking, an optical add-on could be in the form of diffuse optical tomography [31] or fluorescence molecular tomography. Such a hybrid device would either enable better reconstruction by collecting more information about the optical fluence in the sample or measure fluorescence light and hence combine optical absorption contrast with fluorescence contrast. Although optoacoustics can principally measure any absorbing molecules including fluorescent ones, FMT is much more sensitive, especially for low concentrations of fluorescent agents.

On the other hand, an acoustic add-on can provide an additional dimension to optoacoustic tomography, where ultrasound imaging visualizes the anatomical

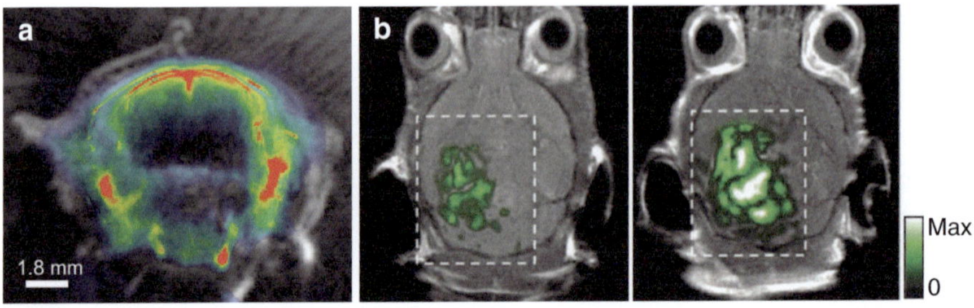


**Fig. 4.6** Multimodal optoacoustic macroscopy (tomography) applications. (**a**) Hybrid optoacoustic (*colors*) and ultrasound (*gray*) imaging of a mouse brain. Courtesy of Ivan Olefir [24]. (**b**) Hybrid optoacoustic (*green*) and MRI (*gray*) imaging of glioblastoma cancer in a mouse brain. Reprinted with permission from [33]

structure of the sample, and optoacoustics gives more information about the molecular and the functional dimensions of a disease or a specific activity [24] (see Fig. 4.6a). Additionally, it is possible to use the ultrasound data for correcting the optoacoustic images by deducing the correct speed of sound from the ultrasound images, which can be later used in the optoacoustic image reconstruction [24, 32]. The beauty of such a hybrid imaging combination is that the ultrasonic detectors (or detector arrays) are readily available and shared by both modalities. Thus, only an ultrasound pulser is needed for the extra ultrasound imaging modality.

Additionally, it is possible to combine the images generated from an MRI modality with those from optoacoustics (see Fig. 4.6b) [33].

### 4.3.6 Conclusion

Optoacoustic tomography is a powerful modality that is capable of whole-body small animal imaging, noninvasively and in vivo. By building upon multiwavelength excitation as enabled by fast and tunable nanosecond lasers, it is possible to add the functional and the molecular dimensions to the captured images and thus to follow not only the anatomy but also the physiology of a disease. So far, this imaging modality has been mostly applied in cancer research and neuroimaging. A strong addition to this modality is the combination of optoacoustics with conventional ultrasound imaging, which is facilitated by commonly employed hardware.

## 4.4 Optoacoustic Mesoscopy

### 4.4.1 Introduction

While optoacoustic macroscopy (tomography) has enabled whole animal or clinical imaging of chromophores at centimeter penetration depths, many biomedical applications, such as imaging microvasculature, early-stage model organisms, or the small-scale spatial heterogeneity of diseases, require higher spatial resolutions than offered by conventional OAT approaches. This demand has led to the development of various high-resolution optoacoustic imaging methods, owing to the inherent scalability of spatial resolution and penetration depth of the optoacoustic modality [1]. More specifically, the detection of optoacoustic frequencies higher than 100 MHz enables spatial resolutions below 30 μm, while the hardware has to be only slightly modified, i.e., high-frequency transducers and laser pulses of only a few nanoseconds have to be used. On the other hand, the penetration depth of high-resolution optoacoustic imaging implementations is fundamentally limited by acoustic attenuation, which acts as a low-pass filter for optoacoustic frequencies [12]. As will be discussed later, in optoacoustic microscopy, spatial resolutions below 1 μm can be realized by using focused illumination, which further reduces the penetration depth to <1 mm.

Optoacoustic mesoscopy seeks to fill the gap between macroscopic and microscopic implementations of the optoacoustic modality by providing high spatial resolutions of several tens of micrometers at intermediate imaging depths of a few millimeters.

### 4.4.2 Scanning Schemes

Most optoacoustic mesoscopy systems are based on a single-element ultrasound transducer that is either cylindrically or spherically focused and that is mechanically scanned for tomographic imaging. As in macroscopy, tomographic image reconstruction algorithms are employed to form three-dimensional volumes of optical absorption in the sample. In the simplest case, a planar scanning geometry is used, i.e., the transducer is scanned laterally within a plane above the sample. Even though optoacoustic mesoscopy has been realized in transmission mode [34], the epi-illumination design, where detection and illumination are located on the same side of the sample, is preferable because it allows for arbitrary specimens to be imaged. One of the first optoacoustic mesoscopy systems (also called acoustic resolution photoacoustic microscopy (AR-PAM)) used a ring-shaped illumination focused into the sample to yield a broad illuminated area, whereas the transducer was positioned in the central dark spot above (see Fig. 4.7a) [35]. Another configuration used a transducer with a central hole that accommodated an optical fiber providing the illumination [36]. A recent implementation developed at the TU Munich is termed raster-scan optoacoustic mesoscopy (RSOM) and relies on a conically shaped spherically focused transducer and several optical fibers guiding the illumination underneath the detector (see Fig. 4.7b) [37]. In all of the aforementioned systems, transducer and illumination are scanned together by means of linear translation stages.

Besides the planar geometry, other scanning schemes have been successfully implemented in the past, such as combinations of cylindrical scanning geometries and linear scans of single detectors (termed MORSOM) [38] or transducer arrays [39, 40] together with a stationary illumination (see Fig. 4.7c). Another system used

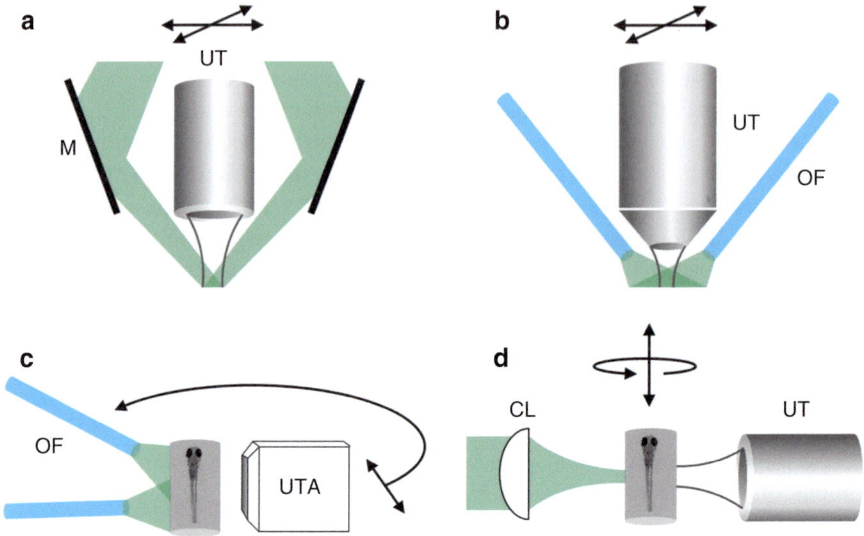


**Fig. 4.7** Optoacoustic mesoscopy implementations. (**a**) Epi-illumination mode dark-field configuration [35]. (**b**) Epi-illumination mode RSOM [37]. (**c**) Hybrid linear and circular scanning configuration [39]. (**d**) Transmission mode light sheet illumination MSOM [41]. Abbreviations: *CL* cylindrical lens, *M* mirror, *OF* optical fiber, *UT* ultrasound transducer, *UTA* ultrasound transducer array. *Green color*: laser illumination

light sheet illumination and cylindrically focused detection in transmission mode (see Fig. 4.7d) [41], in order to reduce out-of-plane signals leading to image artifacts.

### 4.4.3 Spatial Resolution

The spatial resolution of all optoacoustic mesoscopy modalities is governed by the properties of the used ultrasound transducer [42]. The lateral resolution $\delta_{\text{lat}}$ (m) is determined by the width of the acoustic focus and is defined as

$$\delta_{\text{lat}} = 0.71\frac{\lambda_{\text{ac}}}{\text{NA}_{\text{ac}}} \approx 1.4\lambda_{\text{ac}}\frac{F}{D}, \tag{4.15}$$

where $\lambda_{\text{ac}}$ denotes the wavelength of the detected acoustic waves; $F$ and $D$ are the focal distance (m) and the diameter of the active element of the transducer (m), respectively; and $\text{NA}_{\text{ac}}$ represents the numerical aperture of the transducer (unitless) [43]. On the other hand, the axial resolution $\delta_{\text{ax}}$ (m) is given by

$$\delta_{\text{ax}} \approx 0.88\frac{\lambda_{\text{ac}}c}{\text{BW}}, \tag{4.16}$$

where BW is the detection bandwidth of the transducer (Hz).

### 4.4.4 Example Applications

Optoacoustic mesoscopy has been applied to preclinical imaging in numerous studies. The following section provides an overview of several example applications (see Fig. 4.8).

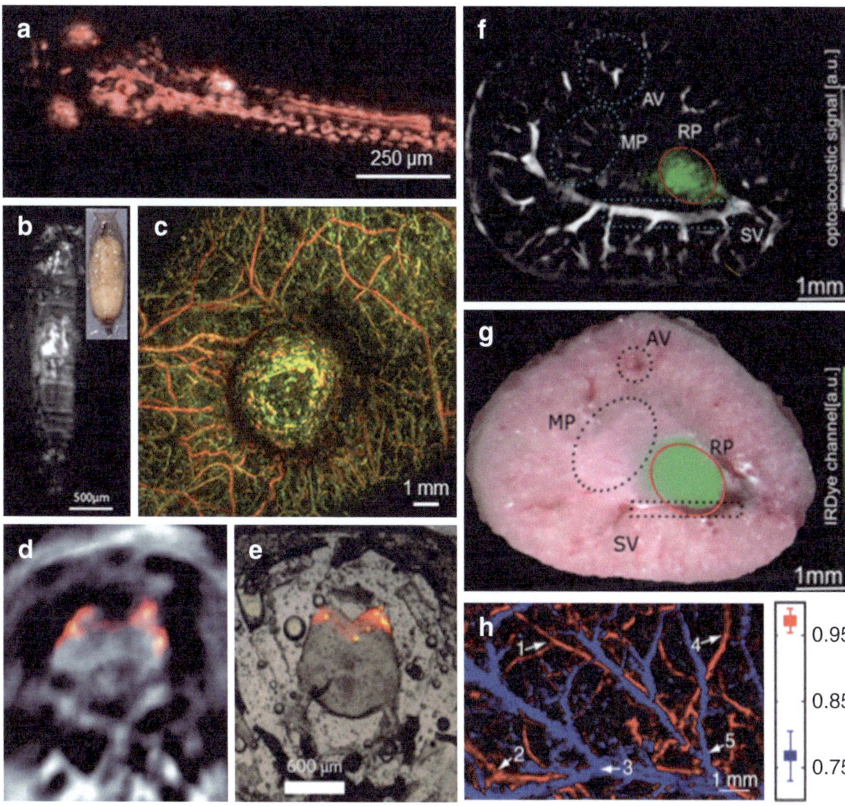


**Fig. 4.8** Imaging applications of optoacoustic mesoscopy. (**a–c**) RSOM imaging of model organisms. (**a**) Imaging of a zebrafish larva ex vivo. *Red color*: low-frequency reconstruction. *White color*: high-frequency reconstruction. Reprinted with permission from [37]. (**b**) Ex vivo imaging of a *Drosophila* pupa, visualizing the anatomical outline and the future wing location via tissue-specific GFP expression. Reprinted with permission from [34]. (**c**) Broadband imaging of a melanoma tumor and the surrounding vascular network in a mouse in vivo. *Red color*: low-frequency reconstruction showing larger blood vessels. *Green color*: high-frequency reconstruction showing smaller features. Reprinted with permission from [44]. (**d, e**) Imaging of mCherry expression in the brain of a juvenile zebrafish in vivo using light sheet illumination MSOM. (**d**) MSOM image showing mCherry signals in color. (**e**) Corresponding histological slice showing fluorescence signals in color. Reprinted with permission from [41]. (**f, g**) Hybrid linear and circular scanning MSOM imaging of an excised mouse kidney. (**f**) Optoacoustic image of the kidney. *Gray color*: high-frequency signals originating from hemoglobin in the sample. *Green color*: low-frequency unmixed signals visualizing the fluorescent dye IRDye 800CW in the renal pelvis (*RP, red circle*). (**g**) Corresponding cryo-section photograph overlaid with the fluorescence signal of the dye (*green color*). Abbreviations: *AV* arcuate vein, *MP* medullary pyramid, *SV* segmental vessel. Reprinted with permission from [39]. (**h**) Dark-field optoacoustic mesoscopy (AR-PAM) imaging of the oxygen saturation ($sO_2$) in a mouse ear in vivo. *Red and blue colors* indicate $sO_2$ levels according to the shown color bar. Reprinted with permission from [35]

1. *Imaging of model organisms*: Model organisms such as a zebrafish (*Danio rerio*), *Drosophila melanogaster*, or *C. elegans* are interesting samples as they are often used by developmental biologists and experimental geneticists to study the development of certain organs (organogenesis), the neuronal system (neurogenesis), or the development of the organisms itself from a single fertile egg (morphogenesis). All these processes or their disruption can be easily monitored in model organisms as they are relatively small, their maintenance is easy, and their reproduction cycle is short. Hence, it is possible to modify certain genes or crossbreed two different lines (e.g., zebrafish), resulting in new generations within a few months. With the RSOM modality, it was possible to image the distribution of melanin-containing cells (melanophores) in an entire zebrafish larva in a label-free manner (see Fig. 4.8a) [37] and to visualize the location of the developing wings in an optically opaque *Drosophila* pupa through labeling with GFP (see Fig. 4.8b) [34]. An improved label-free visualization of zebrafish larvae with isotropic spatial resolution was enabled by MORSOM [38].
2. *Tumor monitoring*: As already pointed out in the discussion of macroscopy applications, optoacoustic imaging is ideally suited to investigate animal models of cancer due to its multispectral capabilities. Beyond that, the higher spatial resolution of optoacoustic mesoscopy allows for the label-free imaging of the vascular network surrounding and supporting tumors. Additionally, the monitoring of vascular growth around the tumor enables an assessment of cancer-related angiogenesis, as has been successfully demonstrated with RSOM in a melanoma mouse model in vivo (Fig. 4.8c) [44]. This capability is an important feature of the optoacoustic modality as it is expected to lead to a more comprehensive understanding of cancer development and related treatment designs.
3. *High-resolution visualization of fluorescent markers in deep tissue*: Multispectral optoacoustic mesoscopy (MSOM) extends the capabilities of MSOT to high spatial resolutions while still achieving millimeters of penetration depth. By using selective plane illumination MSOM, the expression of the fluorescent protein mCherry in the brain of a 6-month-old zebrafish was imaged in vivo (Fig. 4.8d, e) [41]. Such a high-resolution identification of fluorescent reporter molecules in deep tissues is a unique feature of MSOM, overcoming the limitations of fluorescence microscopy in terms of depth or of pure optical tomographic methods in terms of resolution and in vivo applicability.

   Furthermore, MSOM is particularly well suited for the multispectral imaging of the whole organs, as its imaging depth is usually sufficient to penetrate through the entire sample. By using a combination of linear and rotational scanning of two different transducer arrays with distinct detection bandwidths, the distribution of the fluorescent dye IRDye 800CW in the renal pelvis of an excised mouse kidney was imaged together with the vasculature in the kidney (Fig. 4.8f, g) [39].
4. *Visualization of blood oxygen saturation*: The multispectral feature of optoacoustic mesoscopy allows for the separation of oxy- and deoxyhemoglobin and the label-free extraction of metabolic parameters in vivo, such as the blood

oxygen saturation ($sO_2$). One of the first successful demonstrations has been shown on the vascular network in a mouse ear using dark-field optoacoustic mesoscopy (AR-PAM) (Fig. 4.8h) [35].

### 4.4.5 Multimodal Mesoscopic Imaging

Similar to OAT approaches, ultrasound imaging capabilities can be readily integrated in optoacoustic mesoscopy systems, as both modalities share the same transducer. A successful application of hybrid pulse-echo ultrasound and optoacoustic mesoscopic imaging has been demonstrated by the visualization of laser irradiation-induced changes in the cartilage and sclera [36]. The samples were impregnated with bio-functional magnetite nanoparticles for optoacoustic contrast, which penetrated damaged areas of tissue via induced pores and crater-shaped structures. This staining procedure led to a signal increase of damaged areas compared to healthy tissue and allowed for the monitoring of the spatial distribution of laser-induced changes. On the other hand, the ultrasound scans provided mechanical contrast-based information about the changes of tissue thickness induced by the laser irradiation. Figure 4.9 shows the hybrid optoacoustic (top view maximum amplitude projections (MAP), top row) and ultrasound (side view cross sections, bottom row) imaging of sclera tissue ex vivo before (left) and after (right) laser treatment. While both optoacoustic and ultrasound images show structural changes in the sample, the optoacoustic signals are indicative of the formation of pores in the tissue. Such a hybrid imaging approach might lead to promising noninvasive monitoring strategies for laser treatment procedures in ophthalmology and orthopedic applications.

### 4.4.6 Conclusion

Optoacoustic mesoscopy bridges the gap between optoacoustic macroscopy and microscopy applications by providing high spatial resolution (tens of micrometers) at decent imaging depths (several millimeters). It is a promising technique for the noninvasive imaging of the whole organs or small animals. Similar to OAT, optoacoustic mesoscopy offers the opportunity to monitor metabolism or to visualize external contrast agents by capitalizing on the multispectral imaging feature of optoacoustics. So far, multimodal optoacoustic imaging at the mesoscopic scale has been achieved in combination with conventional ultrasound imaging.

## 4.5 Optoacoustic Microscopy

### 4.5.1 Introduction

The development of optical microscopes has revolutionized biomedical research for centuries, as it allows the monitoring of cellular and subcellular processes during

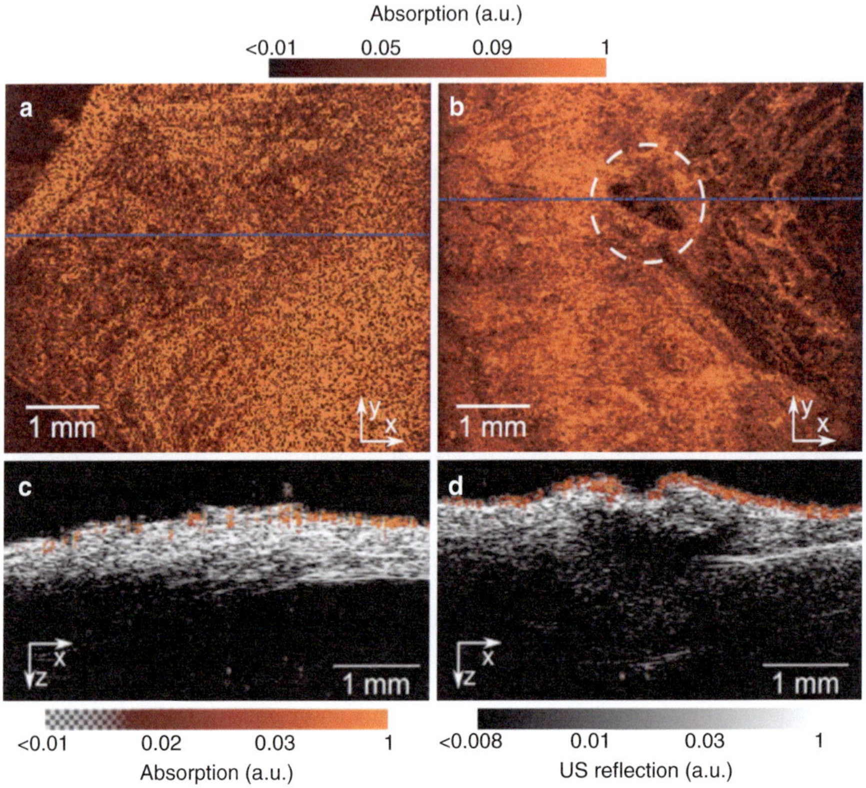


**Fig. 4.9** Hybrid optoacoustic and pulse-echo ultrasound mesoscopic imaging of sclera tissue ex vivo stained with bio-functional magnetite nanoparticles. (**a**, **b**) Optoacoustic top view MAPs of the sclera before (**a**) and after (**b**) laser treatment of the area indicated by the *dashed white circle*. A signal increase due to nanoparticle penetration into laser-induced pores can be observed in (**b**). (**c**, **d**) Side view ultrasound cross sections (*gray*) along the *blue lines* in (**a**, **b**) overlaid with optoacoustic signals (*orange*) before (**c**) and after (**d**) laser treatment. Reprinted with permission from [77]

their occurrence. The contrast of most microscopic systems is based on light scattering, phase differences, or fluorescence. The first two implementations only give morphological information about the specimen, while fluorescence-based contrast requires labeling or staining, which is a cumbersome and lengthy process. In addition to that, although labeled cells might show function as in the case of the fluorescent protein gCamp, the labels might be unstable, weak, and prone to bleaching, and they might lead to phototoxic effects or even cell death.

Optoacoustics, as previously discussed, offers an alternative to the aforementioned optical modalities. By indirectly measuring the absorption of light rather than scattering, phase changes, or fluorescence, it is possible to tap to functional and physiological parameters. Optoacoustic microscopy (OAM, sometimes also termed optical-resolution photoacoustic microscopy—OR-PAM) is based on focused light beams and thus achieves lateral resolutions comparable to traditional optical microscopy techniques, such as confocal or multiphoton microscopy. The generated optoacoustic signals are passively measured using an ultrasound detector. By scanning the focused beam through the sample or by moving the sample itself, it is possible to generate a high-resolution optoacoustic image of optical absorption around the focal plane in the sample.

### 4.5.2 Theory

As previously mentioned, OAM relies on tightly focused optical beams. Therefore, the focusing capability of the microscope defines the lateral resolution, while the axial resolution is still defined through the bandwidth of the acoustic detector. To be more precise, the lateral resolution $\delta_{\mathrm{lat}}$ of OAM is defined as

$$\delta_{\mathrm{lat}} = 0.51\frac{\lambda_{\mathrm{opt}}}{\mathrm{NA}_{\mathrm{opt}}}, \tag{4.17}$$

where $\lambda_{\mathrm{opt}}$ is the wavelength of the excitation light (m) and $\mathrm{NA}_{\mathrm{opt}}$ is the numerical aperture of the focusing lens, which is a number that determines the focusing performance of the optical system [43]. On the other hand, the axial resolution $\delta_{\mathrm{ax}}$ is determined by (4.5) as in the case of optoacoustic mesoscopy. Since focused laser beams are used, the energy requirements are lower in comparison to both optoacoustic mesoscopy and tomography. In typical applications, pulse energies below 100 nJ can be used.

Finally, as in all methods based on optical focusing, the imaging depth is limited by optical scattering to a maximum depth of a few hundred micrometers.

### 4.5.3 Optoacoustic Microscopy Setups

Optoacoustic microscopy setups can be divided into two main categories, as illustrated by Fig. 4.10: transmission and epi-illumination mode implementations. Similar to many optoacoustic mesoscopy applications, in transmission mode, the excitation and the detection are performed on opposite sides of the sample, while in epi-illumination mode, both are performed from the same side.

#### 4.5.3.1 Epi-illumination Mode OAM

Epi-illumination implementations of OAM can be realized in many different ways. One design uses a spherically focused detector with a central hole to accommodate a GRIN lens-tipped fiber, which delivers the focused illumination [36]. Alternatively, it is possible to use transparent detectors such as ring detectors [45, 46] and fiber Bragg gratings (FBGs) [47] or systems similar to the Fabry-Pérot resonator setup introduced in the mesoscopy section [19]. Another method is based on a special kind of transparent acousto-optical beam combiner [48].

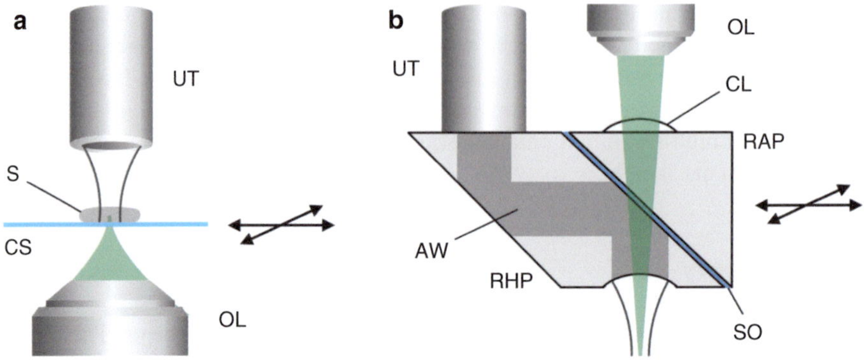


**Fig. 4.10** Optoacoustic microscopy implementations. (**a**) Typical transmission mode configuration using sample scanning [50]. (**b**) Epi-illumination mode OAM based on a mechanically scanned acousto-optical beam combiner (*RAP and RHP*) [78]. Abbreviations: *AW* acoustic waves, *CL* correction lens, *CS* cover slip, *OL* objective lens, *RAP* right angle prism, *RHP* rhomboid prism, *S* sample, *SO* silicone oil, *UT* ultrasound transducer. *Green color*: laser illumination

The key advantage of epi-illumination mode OAM over transmission mode designs is the applicability to samples with arbitrary geometries. On the other hand, many epi-illumination mode setups are limited in spatial resolution, because the space demands for the acoustic and optical components typically impede the usage of high NA objectives.

#### 4.5.3.2 Transmission Mode OAM

In common transmission mode configurations, the acoustic detector and the focused optical illumination are coaxially aligned. Furthermore, the detector and the excitation can be either confocally aligned in order to increase the sensitivity, or the acoustic detector can be slightly defocused in order to increase the scanning field of view (FOV). The latter design leads to an increased imaging speed by rapidly scanning the laser beam within the acoustic sensitivity field by means of scanning mirrors or other beam deflection components. In the system developed at the TU Munich, the first approach has been pursued, where both the acoustic detector and the optical focusing are aligned in a confocal manner. The sample is placed on a fully motorized xyz stage, and a three-dimensional scan is performed. The scanning head is mounted on top of an inverted multimodal microscope, which will be described in more detail in the next section. Additionally, several detectors can be used and are easily interchangeable. The advantage of a transmission mode design over epi-illumination mode is the easiness of alignment and the possibility to use any kind of ultrasound detectors [49, 50]. Another advantage over the epi-illumination mode design is the variety of compatible optics, where it is possible to use the same high-quality and high NA objectives as employed in fluorescence microscopy. Hence, it is possible to easily combine OAM with other optical microscopy techniques, such as confocal or multiphoton microscopy, as we will see in the next section.

The system of the TU Munich is based on a commercial microscope where instead of a CW laser commonly used in fluorescence microscopy, a pulsed nanosecond laser is coupled into the objective. The laser and ultrasound detector are confocally aligned for maximum sensitivity, and an image is generated by scanning the confocal spot through the sample.

In another implementation, the laser beam is rapidly raster-scanned in the sample by means of a set of galvanometric mirrors. For this purpose, the ultrasound detector is lifted, and the beam is scanned in the positive defocus of the acoustic sensitivity field. Such a configuration allows for FOVs of ~600 × 600 $\mu m^2$ to be scanned within a few minutes while achieving sufficient sensitivity. For even larger fields of view, a combination of mechanical and galvanometric scanning could be used [51, 52].

### 4.5.4 Example Applications

Figure 4.11 provides a summary of exemplary OAM applications. Aside from the label-free imaging of single red blood cells, melanophores, or melanoma cells (see Fig. 4.11a for the latter [53]), OAM has been applied in the following fields (for more information, the reader is referred to the references mentioned in this section):

1. *Imaging of model organisms*: As previously discussed, model organisms are widely used in biology to study the development of organs or the organisms themselves. In the case of the group of the TU Munich, melanophores forming the pigmentation of zebrafish larvae were imaged with subcellular spatial resolution using a transmission mode OAM system and an optical sensor based on a π-shifted fiber Bragg grating [47] (see Fig. 4.11b). For a more isotropic spatial resolution and thus better visualization results regarding structures oriented parallel to the detection axis, multi-view methods similar to mesoscopic MORSOM can be used [54].
2. *Mouse ear vasculature visualization*: Mouse ears are interesting organs as they contain a dense network of vasculature and microvasculature and are additionally thin in nature (typically around 300 μm). Thus, they are readily accessible by transmission mode setups, where the ear is flattened between the illumination and the ultrasonic detector. On the one hand, mouse ears are frequently used for highlighting the capabilities of OAM systems in imaging a range of different vascular sizes. On the other hand, similar to any part of the animal, it is possible to study a multitude of different diseases in a mouse ear [55], such as cancer growth [43].
3. *Neuroimaging*: The oxygenation state of a vascular network is linked to the activity of the organ fed by these vessels. This is because blood carries both nutrients and oxygen necessary for metabolic processes, which increase during organ activity. BOLD-MRI is capitalizing on the same principle when monitoring the neuro-activity in a human brain. Similar to MRI, as we described in

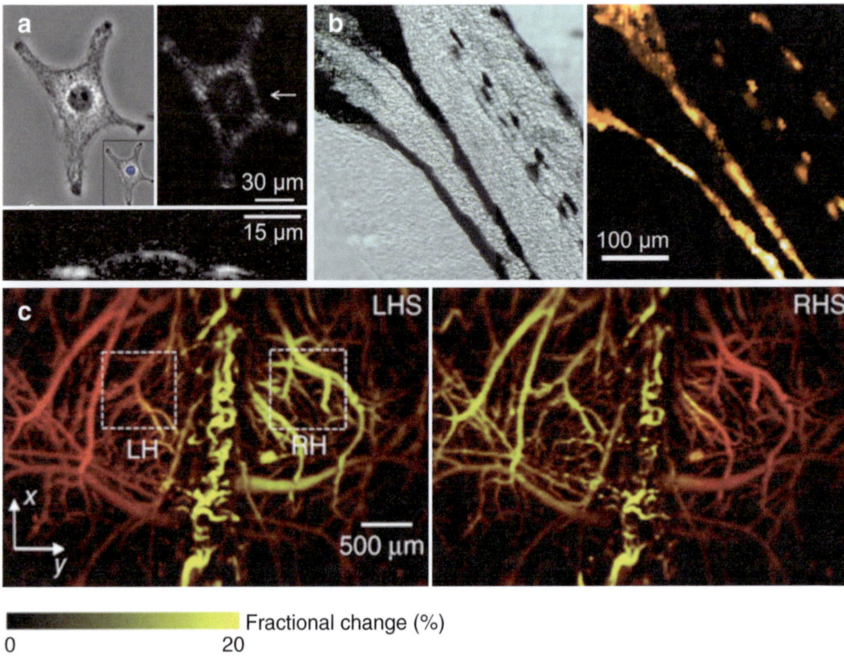


**Fig. 4.11** Examples of optoacoustic microscopy applications. (**a**) Optoacoustic imaging of a single melanoma cell. *Upper left*: bright-field image. *Upper right*, OAM top view MAP image; *lower*, OAM side view MAP image, reprinted with permission from [79]. (**b**) Bright-field (*left*) and corresponding optoacoustic microscopy (*right*) imaging of melanophores in a zebrafish trunk ex vivo. Courtesy of Georg Wissmeyer and Dominik Soliman. (**c**) Optoacoustic microscopy imaging of neuro-activity in a mouse brain in vivo during the stimulation of the left (*LHS*) and right (*RHS*) hind limb. Abbreviations: *LH* left hemisphere, *RH* right hemisphere. Reprinted with permission from [51]

the section about optoacoustic macroscopy, optoacoustics is capable of monitoring the oxygen level of blood in real-time, noninvasively, and in a label-free manner. In contrast to optoacoustic tomography, OAM is able to capture functional and morphological changes in microvasculature. Hence, it is possible not only to monitor the global behavior but also local changes on a microscopic scale. As in OAT, this activity can be recorded by using multiple excitation wavelengths, which are appropriately chosen to separate both the oxy and the deoxy components of the blood. Additionally, because higher light fluence rates are possible in microscopy applications, it is possible to take advantage of nonlinear optoacoustic effects such as absorption saturation. Based on the difference in the absorption relaxation rate, oxy- and deoxyhemoglobin can be separated, allowing for the determination of oxygen saturation levels. This method has been used to monitor neuronal activity in a mouse brain in vivo (see Fig. 4.11c [51]).

### 4.5.5 Multimodal Microscopic Imaging

As OAM uses similar components as other optical microscopy systems, it can be readily combined with other high-resolution modalities within a common hybrid imaging framework. In the case of the system at the TU Munich, the OAM system is built around a commercial inverted microscopy stand and is combined with multiphoton microscopy, incorporating second harmonic generation (SHG), third harmonic generation (THG), and two-photon excitation fluorescence (TPEF) microscopy [49, 50]. Different laser sources are used for the different modalities, which are coupled into the same microscope. For OAM, a nanosecond laser is used, while multiphoton microscopy is based on a femtosecond laser to excite nonlinear optical effects. The integration of all these modalities into a common device enables the hybrid imaging of the same sample without the need of moving it between different systems. At the same time, an enrichment of accessible label-free contrast mechanisms is achieved, which enables the concurrent visualization of different anatomical features in biological specimens and other samples. Figure 4.12a shows an example of the label-free hybrid imaging of a zebrafish tail ex vivo, where muscles, connective tissue, and melanophores are simultaneously visualized through SHG (blue), THG (green), and OAM (red) imaging, respectively [50].

In another application, the necrotic core region of a human carotid atheroma slice was imaged with the hybrid microscope (see Fig. 4.12b). The SHG (green), THG (blue), TPEF (yellow), and OAM (red) signals visualize collagen, tissue morphology, elastin, and blood embeddings, respectively. This multimodal and staining-free imaging method is expected to play an important role in understanding how plaques develop and thus devising the right strategies for treatment [52]. Similarly, such a technology could be used for better understanding of cancer development in model animals and the mutual interactions of different molecules inside the tumors.

Figure 4.12c shows the hybrid OAM and multiphoton imaging of a mouse ear in vivo, visualizing blood vessels through OAM (red), cell boundaries and hair follicles through THG (blue), as well as collagen via SHG (green) imaging in 3D [56].

In other applications, OAM has been successfully combined with fluorescence [57] and confocal microscopy [58], ultrasound imaging [59], or optical coherence tomography (OCT) [60].

### 4.5.6 Conclusion

Optoacoustic microscopy is a potent technology that allows for the visualization of optical absorption, related to different functional processes and disease biomarkers, at a microscopic scale. In this section, we discussed different implementations as well as various applications of OAM and gave a more detailed discussion about the system developed at the TU Munich. This setup, although based on a transmission

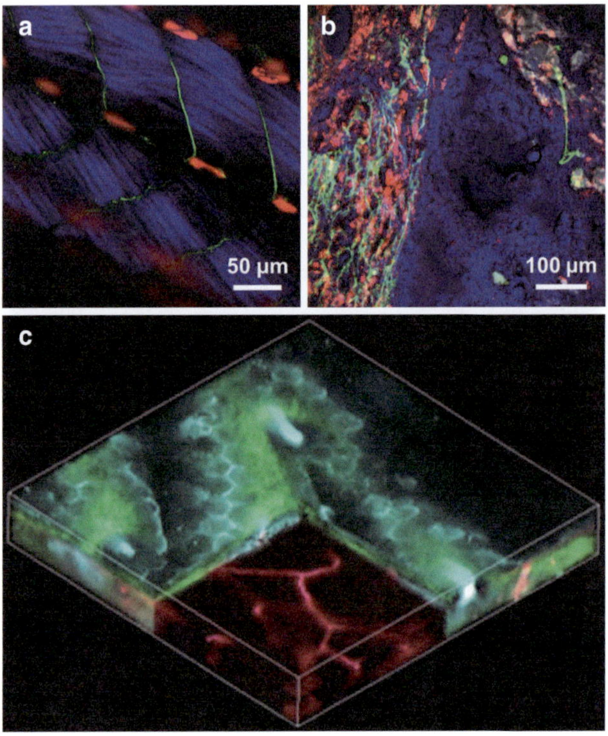


**Fig. 4.12** Multimodal optoacoustic and multiphoton microscopy applications. (**a**) Hybrid imaging of a zebrafish trunk ex vivo, visualizing the musculature (SHG, *blue*), connective tissue (THG, *green*), and melanophores (OAM, *red*). Reprinted with permission from [50]. (**b**) Hybrid imaging of the necrotic core region of a human carotid atheroma slice, visualizing collagen fibrils (SHG, *green*), elastin (TPEF, *yellow*), overall tissue morphology (THG, *blue*), and blood embeddings (OAM, *red*). Reprinted with permission from [52]. (**c**) Hybrid 3D imaging of a mouse ear in vivo, visualizing collagen (SHG, *green*), hair follicles and cell boundaries (THG, *blue*), as well as blood vessels (OAM, *red*). Reprinted with permission from [56]

mode configuration, allows for the imaging of a multitude of different samples, ranging from model organisms, such as zebrafish and mouse ears, to tissue slices and cell cultures. Limitations of OAM include its reliance on optical focusing, which does not allow for penetration beyond a few hundred micrometers. On the other hand, this technology enables the label-free imaging of disease biomarkers and other bio-components that cannot be imaged otherwise or require potentially harmful labeling or staining procedures. Because OAM uses similar laser sources and optical components as other laser microscopy techniques, it is relatively simple to combine OAM with other optical microscopic modalities, such as multiphoton and confocal microscopy or OCT, in order to simultaneously gather comprehensive anatomical information from complex biological samples. Furthermore, similar to the macroscopic and mesoscopic case, OAM can be combined with pulse-echo ultrasound imaging without the need for complicated additional hardware.

## 4.6 Summary and Outlook

In this chapter, we introduced optical and optoacoustic imaging as well as the combination of these techniques with other modalities, such as ultrasound imaging and various optical imaging technologies. Because of advances in laser and ultrasonic detection technology, optoacoustics has developed into a strong imaging modality over the last two decades and has been applied in multiple applications, such as neuroimaging [24, 51], cancer research [44, 61], imaging of the bio-distribution of certain molecules, as well as cardiovascular imaging [22]. The versatility of optoacoustics is facilitated by its inherent scalability, enabling macroscopic, mesoscopic, and microscopic imaging applications based on similar technology.

In the future, we expect an increasing number of cases where optoacoustics becomes a mainstream imaging method, especially when combined with other modalities (i.e., multimodal optoacoustic imaging). In such cases, optoacoustics is complemented with modalities that either give a different view on the anatomy of the examined specimen or a more sensitive measurement of fluorescently labeled molecular agents. Additionally, by developing biomolecules that are engineered toward yielding stronger optoacoustic signals [29, 62–64], optoacoustic imaging is likely to become a more rounded technology on its own.

Another potential future mainstream direction is clinical imaging, where the same technologies described in this chapter could be used in a clinical setting, either for early and metastatic cancer diagnosis [65–67], endoscopic applications [68, 69], tissue oxygenation [70], intravascular imaging [71], dermatology [65, 72, 73], or other applications [74–76].

## References

1. Ntziachristos V. Going deeper than microscopy: the optical imaging frontier in biology. Nat Methods. 2010;7(8):603–14.
2. Ale A, Ermolayev V, Herzog E, Cohrs C, de Angelis MH, Ntziachristos V. FMT-XCT: in vivo animal studies with hybrid fluorescence molecular tomography–X-ray computed tomography. Nat Methods. 2012;9(6):615–20.
3. Ntziachristos V, Yodh AG, Schnall M, Chance B. Concurrent MRI and diffuse optical tomography of breast after indocyanine green enhancement. Proc Natl Acad Sci U S A. 2000;97(6):2767–72.
4. Bowen T. Radiation-induced thermoacoustic soft tissue imaging. In: IEEE Ultrasonic Symposium; 1981.
5. Wang LV, Wu H-I. Biomedical optics: principles and imaging. Hoboken, NJ: Wiley Interscience; 2007.
6. Diebold GJ, Sun T, Khan MI. Photoacoustic monopole radiation in one, two, and three dimensions. Phys Rev Lett. 1991;67(24):3384–7.
7. Westervelt PJ, Larson RS. Laser-excited broadside array. J Acoust Soc Am. 1973;54(1):121–2.
8. Bossy E, Gigan S. Photoacoustics with coherent light. Photoacoustics. 2016;4(1):22–35.
9. Khan MI, Diebold GJ. The photoacoustic effect generated by an isotropic solid sphere. Ultrasonics. 1995;33(4):265–9.
10. Khan MI, Diebold GJ. The photoacoustic effect generated by laser irradiation of an isotropic solid cylinder. Ultrasonics. 1996;34(1):19–24.

11. Treeby BE, Cox BT, Zhang EZ, Patch SK, Beard PC. Measurement of broadband temperature-dependent ultrasonic attenuation and dispersion using photoacoustics. IEEE Trans Ultrason Ferroelectr Freq Control. 2009;56(8):1666–76.
12. Deán-Ben XL, Razansky D, Ntziachristos V. The effects of acoustic attenuation in optoacoustic signals. Phys Med Biol. 2011;56(18):6129–48.
13. Szabo TL. Diagnostic ultrasound imaging: inside out. Amsterdam: Academic Press; 2004.
14. Rosenthal A, Ntziachristos V, Razansky D. Acoustic inversion in optoacoustic tomography: a review. Curr Med Imaging Rev. 2013;9:318–36.
15. Xu M, Wang LV. Universal back-projection algorithm for photoacoustic computed tomography. Phys Rev E. 2005;71:016706.
16. Tomography—Wikipedia [Online]. Available: https://en.wikipedia.org/wiki/Tomography. Accessed 4 Sep 2016.
17. Buehler A, Kacprowicz M, Taruttis A, Ntziachristos V. Real-time handheld multispectral optoacoustic imaging. Opt Lett. 2013;38(9):1404–6.
18. Deán-Ben XL, Razansky D. Adding fifth dimension to optoacoustic imaging: volumetric time-resolved spectrally enriched tomography. Light Sci Appl. 2014;3(1):e137.
19. Laufer J, Johnson P, Zhang E, Treeby B, Cox B, Pedley B, Beard P. In vivo preclinical photoacoustic imaging of tumor vasculature development and therapy. J Biomed Opt. 2012;17(5):56016.
20. Herzog E, Taruttis A, Beziere N, Lutich AA, Razansky D. Optical imaging of cancer heterogeneity with multispectral. Radiology. 2012;263(2):461–8.
21. Beziere N, Schacky C, Kosanke Y, Kimm M, Nunes A, Licha K, Aichler M, Walch A, Rummeny EJ, Ntziachristos V, et al. Optoacoustic imaging and staging of inflammation in a murine model of arthritis. Arthritis Rheumatol. 2014;66(8):2071–8.
22. Taruttis A, Wildgruber M, Kosanke K, Beziere N, Licha K, Haag R, Aichler M, Walch A, Rummeny E, Ntziachristos V. Multispectral optoacoustic tomography of myocardial infarction. Photoacoustics. 2013;1(1):3–8.
23. Taruttis A, Ntziachristos V. Advances in real-time multispectral optoacoustic imaging and its applications. Nat Photon. 2015;9(4):219–27.
24. Olefir I, Merčep E, Burton NC, Ovsepian SV, Ntziachristos V. Hybrid multispectral optoacoustic and ultrasound tomography for morphological and physiological brain imaging. J Biomed Opt. 2016;21(8):86005.
25. Nasiriavanaki M, Xia J, Wan H, Bauer AQ, Culver JP, Wang LV. High-resolution photoacoustic tomography of resting-state functional connectivity in the mouse brain. Proc Natl Acad Sci U S A. 2013;8943:9–13.
26. Gottschalk S, Fehm TF, Deán-Ben XL, Tsytsarev V, Razansky D. Correlation between volumetric oxygenation responses and electrophysiology identifies deep thalamocortical activity during epileptic seizures. Neurophotonics. 2017;4(1):11007.
27. Tzoumas S, Nunes A, Olefir I, Stangl S, Symvoulidis P, Glasl S, Bayer C, Multhoff G, Ntziachristos V. Eigenspectra optoacoustic tomography achieves quantitative blood oxygenation imaging deep in tissues. Nat Commun. 2016;7:12121.
28. Bohndiek SE, Sasportas LS, Machtaler S, Jokerst JV, Hori S, Gambhir SS. Photoacoustic tomography detects early vessel regression and normalization during ovarian tumor response to the antiangiogenic therapy trebananib. J Nucl Med. 2015;56(12):1942–7.
29. Jathoul AP, Laufer J, Ogunlade O, Treeby B, Cox B, Zhang E, Johnson P, Pizzey AR, Philip B, Marafioti T, et al. Deep in vivo photoacoustic imaging of mammalian tissues using a tyrosinase-based genetic reporter. Nat Photon. 2015;9:239–46.
30. Razansky D, Buehler A, Ntziachristos V. Volumetric real-time multispectral optoacoustic tomography of biomarkers. Nat Protoc. 2011;6(8):1121–9.
31. Bauer AQ, Nothdurft RE, Erpelding TN, Wang LV, Culver JP. Quantitative photoacoustic imaging: correcting for heterogeneous light fluence distributions using diffuse optical tomography. J Biomed Opt. 2011;16(9):96016.
32. Mercep E, Burton NC, Claussen J, Razansky D. Whole-body live mouse imaging by hybrid reflection-mode ultrasound and optoacoustic tomography. Opt Lett. 2015;40(20):4643–6.

33. Kircher M, La Zerda AD, Jokerst J, Zavaleta C. A brain tumor molecular imaging strategy using a new triple-modality MRI-photoacoustic-Raman nanoparticle. Nat Med. 2012;18:829–34.
34. Omar M, Gateau J, Ntziachristos V. Raster-scan optoacoustic mesoscopy in the 25–125 MHz range. Opt Lett. 2013;38(14):2472–4.
35. Zhang HF, Maslov K, Stoica G, Wang LV. Functional photoacoustic microscopy for high-resolution and noninvasive in vivo imaging. Nat Biotechnol. 2006;24(7):848–51.
36. Estrada H, Turner J, Kneipp M, Razansky D. Real-time optoacoustic brain microscopy with hybrid optical and acoustic resolution. Laser Phys Lett. 2014;11(4):45601.
37. Omar M, Soliman D, Gateau J, Ntziachristos V. Ultrawideband reflection-mode optoacoustic mesoscopy. Opt Lett. 2014;39(13):3911–4.
38. Omar M, Rebling J, Wicker K, Schmitt-Manderbach T, Schwarz M, Gateau J, López-Schier H, Mappes T, Ntziachristos V. Optical imaging of post-embryonic zebrafish using multi orientation raster scan optoacoustic mesoscopy. Light Sci Appl. 2017;6:e16186.
39. Chekkoury A, Gateau J, Driessen W, Symvoulidis P, Bézière N, Feuchtinger A, Walch A, Ntziachristos V. Optical mesoscopy without the scatter: broadband multispectral optoacoustic mesoscopy. Biomed Opt Express. 2015;6(9):3134–48.
40. Chekkoury A, Nunes A, Gateau J, Symvoulidis P, Feuchtinger A, Beziere N, Ovsepian SV, Walch A, Ntziachristos V. High-resolution multispectral optoacoustic tomography of the vascularization and constitutive hypoxemia of cancerous tumors. Neoplasia. 2016;18(8):459–67.
41. Razansky D, Distel M, Vinegoni C, Ma R, Perrimon N, Köster RW, Ntziachristos V. Multispectral opto-acoustic tomography of deep-seated fluorescent proteins in vivo. Nat Photon. 2009;3(7):412–7.
42. Cobbold RSC. Foundations of biomedical ultrasound. New York: Oxford University Press; 2006.
43. Yao J, Wang LV. Photoacoustic microscopy. Laser Photon Rev. 2014;7(5):1–36.
44. Omar M, Schwarz M, Soliman D, Symvoulidis P, Ntziachristos V. Pushing the optical imaging limits of cancer with multi-frequency-band raster-scan optoacoustic mesoscopy (RSOM). Neoplasia. 2015;17(2):208–14.
45. Li H, Dong B, Zhang Z, Zhang HF, Sun C. A transparent broadband ultrasonic detector based on an optical micro-ring resonator for photoacoustic microscopy. Sci Rep. 2014;4:4496.
46. Huang S-W, Chen S-L, Ling T, Maxwell A, O'Donnell M, Guo LJ, Ashkenazi S. Low-noise wideband ultrasound detection using polymer microring resonators. Appl Phys Lett. 2008;92(19):193509.
47. Wissmeyer G, Soliman D, Shnaiderman R, Rosenthal A, Ntziachristos V. All-optical optoacoustic microscope based on wideband pulse interferometry. Opt Lett. 2016;41(9):1953–6.
48. Maslov K, Zhang HF, Hu S, Wang LV. Optical-resolution photoacoustic microscopy for in vivo imaging of single capillaries. Opt Lett. 2008;33(9):929–31.
49. Tserevelakis GJ, Soliman D, Omar M, Ntziachristos V. Hybrid multiphoton and optoacoustic microscope. Opt Lett. 2014;39(7):1819–22.
50. Soliman D, Tserevelakis GJ, Omar M, Ntziachristos V. Combining microscopy with mesoscopy using optical and optoacoustic label-free modes. Sci Rep. 2015;5:12902.
51. Yao J, Wang L, Yang J-M, Maslov KI, Wong TTW, Li L, Huang C-H, Zou J, Wang LV. High-speed label-free functional photoacoustic microscopy of mouse brain in action. Nat Methods. 2015;12(5):407–10.
52. Seeger M, Karlas A, Soliman D, Pelisek J, Ntziachristos V. Multimodal optoacoustic and multiphoton microscopy of human carotid atheroma. Photoacoustics. 2016;4:102–11.
53. Shelton RL, Mattison SP, Applegate BE. Volumetric imaging of erythrocytes using label-free multiphoton photoacoustic microscopy. J Biophoton. 2014;7(10):834–40.
54. Zhu L, Li L, Gao L, Wang LV. Multiview optical resolution photoacoustic microscopy. Optica. 2014;1(4):217–22.
55. Wang L, Maslov K, Yao J, Rao B, Wang LV. Fast voice-coil scanning optical-resolution photoacoustic microscopy. Opt Lett. 2011;36(2):139–41.

56. Song W, Xu Q, Zhang Y, Zhan Y, Zheng W, Song L. Fully integrated reflection-mode photoacoustic, two-photon, and second harmonic generation microscopy in vivo. Sci Rep. 2016;6:32240.
57. Tserevelakis GJ, Tsagkaraki M, Zacharakis G. Hybrid photoacoustic and optical imaging of pigments in vegetative tissues. J Microsc. 2016;263:300–6.
58. Wang Y, Maslov K, Kim C, Hu S, Wang LV. Integrated photoacoustic and fluorescence confocal microscopy. IEEE Trans Biomed Eng. 2010;57(10):2576–8.
59. Harrison T, Ranasinghesagara JC, Lu H, Mathewson K, Walsh A, Zemp RJ. Combined photoacoustic and ultrasound biomicroscopy. Opt. Express. 2009;17(24):22041–6.
60. Jiao S, Xie Z, Zhang HF, Puliafito CA. Simultaneous multimodal imaging with integrated photoacoustic microscopy and optical coherence tomography. Opt Lett. 2009;34(19):2961–3.
61. Yao J, Shcherbakova DM, Li C, Krumholz A, Lorca RA, Reinl E, England SK, Verkhusha VV, Wang LV. Reversibly switchable fluorescence microscopy with enhanced resolution and image contrast. J Biomed Opt. 2014;19:86018.
62. Stiel AC, Deán-Ben XL, Jiang Y, Ntziachristos V, Razansky D, Westmeyer GG. High-contrast imaging of reversibly switchable fluorescent proteins via temporally unmixed multispectral optoacoustic tomography. Opt Lett. 2015;40(3):367–70.
63. Jiang Y, Sigmund F, Reber J, Deán-Ben XL, Glasl S, Kneipp M, Estrada H, Razansky D, Ntziachristos V, Westmeyer GG. Violacein as a genetically-controlled, enzymatically amplified and photobleaching-resistant chromophore for optoacoustic bacterial imaging. Sci Rep. 2015;5:11048.
64. Stritzker J, Kirscher L, Scadeng M, Deliolanis NC, Morscher S, Symvoulidis P, Schaefer K, Zhang Q, Buckel L, Hess M, et al. Vaccinia virus-mediated melanin production allows MR and optoacoustic deep tissue imaging and laser-induced thermotherapy of cancer. Proc Natl Acad Sci. 2013;110(9):3316–20.
65. Aguirre J, Schwarz M, Soliman D, Buehler A, Omar M, Ntziachristos V. Broadband mesoscopic optoacoustic tomography reveals skin layers. Opt Lett. 2014;39(21):6297.
66. Schwarz M, Omar M, Buehler A, Aguirre J, Ntziachristos V. Implications of ultrasound frequency in optoacoustic mesoscopy of the skin. IEEE Trans Med Imaging. 2015;34(2):672–7.
67. Stoffels I, Morscher S, Helfrich I, Hillen U, Leyh J, Burton NC, Sardella TCP, Claussen J, Poeppel TD, Bachmann HS, et al. Metastatic status of sentinel lymph nodes in melanoma determined noninvasively with multispectral optoacoustic imaging. Sci Transl Med. 2015;7(317):317ra199.
68. Yang J-M, Favazza C, Chen R, Yao J, Cai X, Maslov K, Zhou Q, Shung KK, Wang LV. Simultaneous functional photoacoustic and ultrasonic endoscopy of internal organs in vivo. Nat Med. 2012;18(8):1297–302.
69. He H, Buehler A, Ntziachristos V. Optoacoustic endoscopy with curved scanning. Opt Lett. 2015;40(20):4667–70.
70. Diot G, Dima A, Ntziachristos V. Multispectral opto-acoustic tomography of exercised muscle oxygenation. Opt Lett. 2015;40(7):1496–9.
71. Sethuraman S, Amirian JH, Litovsky SH, Smalling RW, Emelianov SY. Spectroscopic intravascular photoacoustic imaging to differentiate atherosclerotic plaques. Opt Express. 2008;16(5):3362–7.
72. Aguirre J, Schwarz M, Garzorz N, Omar M, Buehler A, Eyerich K, Ntziachristos V. Precision assessment of label-free psoriasis biomarkers with ultra-broadband optoacoustic mesoscopy. Nat Biomed Eng. 2017;1:0068.
73. Schwarz M, Soliman D, Omar M, Buehler A, Aguirre J, Ntziachristos V. Optoacoustic dermoscopy of the human skin: tuning excitation energy for optimal detection bandwidth with fast and deep imaging in vivo. IEEE Trans Med Imaging. 2017;36:1287–96.
74. Taruttis A, Timmermans AC, Wouters PC, Kacprowicz M, van Dam GM, Ntziachristos V. Optoacoustic imaging of human vasculature: feasibility by using a handheld probe. Radiology. 2016;281(1):256–63.

75. Neuschmelting V, Burton NC, Lockau H, Urich A, Harmsen S, Ntziachristos V, Kircher MF. Performance of a multispectral optoacoustic tomography (MSOT) system equipped with 2D vs. 3D handheld probes for potential clinical translation. Photoacoustics. 2016;4(1):1–10.
76. Waldner MJ, Knieling F, Egger C, Morscher S, Claussen J, Vetter M, Kielisch C, Fischer S, Pfeifer L, Hagel A, et al. Multispectral optoacoustic tomography in Crohn's disease: non-invasive imaging of disease activity. Gastroenterology. 2016;151:238–40.
77. Estrada H, Sobol E, Baum O, Razansky D. Hybrid optoacoustic and ultrasound biomicroscopy monitors' laser-induced tissue modifications and magnetite nanoparticle impregnation. Laser Phys Lett. 2014;11(12):125601.
78. Hu S, Maslov K, Wang LV. Second-generation optical-resolution photoacoustic microscopy with improved sensitivity and speed. Opt Lett. 2011;36(7):1134–6.
79. Strohm EM, Berndl ESL, Kolios MC. High frequency label-free photoacoustic microscopy of single cells. Photoacoustics. 2013;1(3):49–53.

# Small Animal Imaging in Oncology Drug Development

# 5

Joseph D. Kalen and James L. Tatum

## 5.1 Introduction

Major advances in small animal imaging have been made during the last two decades encompassing a full array of platforms that image along the electromagnetic spectrum from MRI ($10^0$–$10^1$ m), optical ($10^{-6}$ m), X-ray ($10^{-9}$ m), to nuclear ($10^{-11}$–$10^{-12}$ m). This in part has been facilitated by the National Cancer Institute (NCI), National Institutes of Health (NIH) through the support of Small Animal Imaging Research Programs (SAIRP), and other initiatives to increase the availability of small animal imaging platforms and develop the expertise in the use of these methods. While the primary application of these new techniques has been research tools to answer scientific questions especially related to the understanding of in vivo systems, another area of interest has been the introduction of imaging-based in vivo assay systems for drug development in oncology. In fact, a major effort has been undertaken to integrate in vivo imaging biomarker development with in vitro biomarker development in contrast to the historical scenario of applying imaging only late in the development plan, leading to the conundrum of validation of imaging while trying to employ imaging as a biomarker.

Drug development is a high-risk business in which late-stage failures are especially costly, with an average cost (capitalized and out of pocket) (2016) of

J. D. Kalen (✉)
Small Animal Imaging Program, Laboratory Animal Sciences Program, Frederick National Laboratory for Cancer Research Sponsored by the National Cancer Institute, Frederick, MD, USA
e-mail: kalenj@mail.nih.gov

J. L. Tatum
Cancer Imaging Program, Division of Cancer Treatment and Diagnosis, National Cancer Institute, NIH, Bethesda, MD, USA
e-mail: james.tatum@nih.gov



C. Kuntner-Hannes, Y. Haemisch (eds.), *Image Fusion in Preclinical Applications*, https://doi.org/10.1007/978-3-030-02973-9_5

approximately \$3.95B (\$1.528B (preclinical) and \$2.43B (clinical)) normalized to 2013 dollars [1]. Trends in capitalized costs since the 1970s (based on 2013 dollars) have shown dramatic increases in preclinical and clinical costs, 1007% and 2085%, respectively [1]. Although the monetary costs are obvious, the opportunity cost of such failures, consuming valuable resources and time, may lead to more significant health cost. The focus on targeted agents has further complicated the development process. Unfortunately, late-phase failures with new targeted agents are not uncommon and frequently the result of inadequate biomarker development. It has been reported that robust biomarkers are essential to successful drug development and can improve the success rate for phase I to drug approval as much as 25.9% (with biomarkers) as compared to 8.4% (without biomarkers) [2]. In oncology, drugs and especially new pathway-specific drugs non-context in vitro assays may not be adequate when applied to clinical scenarios which are highly contextually based. Small animal imaging with image fusion encompassing different modalities and molecular probes has the potential to enhance our understanding of drug candidates and combinations in context by serving as an in vivo assay system allowing evaluation of various biomarkers in the complex biologic system of cancer. While imaging of small animals is now routinely performed daily in many labs, converting such imaging to an in vivo assay for drug development is significantly more challenging.

Performing small animal imaging in the context of an in vivo assay for drug development encompasses numerous aspects: standardization of equipment processes (SOPs) including quality control and quality assurance, data acquisition, analysis, and validation of output with respect to a gold standard (i.e., pathology) [3–6] are all important requirements for developing an assay for drug development. Furthermore, other aspects such as animal handling, anesthesia and understanding its effect on the animals physiology (pulmonary and cardiac functions and stress levels) [7–11], personnel safety for handling animals that contain toxic chemicals [12], validation of animal model platforms [5], multi-animal throughput to obtain statistical significance, and multi-imaging platforms to discern a drugs effect on the various physiological [13], anatomical, and molecular pathways are all significant aspects for developing a robust drug development in vivo assay. Numerous references have been cited that describe the above factors for performing small animal imaging.

The main aspect of this book is to provide an overview of multimodality preclinical imaging, and this chapter is intended to demonstrate how these techniques can be implemented for oncology drug development. Initial utilization of in vivo imaging incorporated clinical scanners for simple analysis of anatomical tumor volumes or a metabolic function using the clinically available radiopharmaceutical 2-deoxy-2-($^{18}$F)fluoro-D-glucose [$^{18}$F]FDG. As equipment manufacturers developed and modified scanners explicitly for small animals, incorporating higher spatial image resolution and faster acquisitions to analyze the small animal's rapid heart, pulmonary, and biological rates, these scanners were providing in vivo techniques to allow researchers the ability to question and comprehend specific processes pertinent to developing an oncologic drug. Furthermore, due to the inherent physics of each system's attributes to acquire data along the electromagnetic spectrum and

**Table 5.1** Comparison of preclinical in vivo imaging modalities

| Modality | Image resolution (μm) | 3D capability (tomography) | Probe sensitivity (mol/L) | Intrinsic contrast | Amount of probe required | Activatable probes | Dynamic studies |
|---|---|---|---|---|---|---|---|
| X-ray CT | 5 | Yes | N/A | Yes | | No | No |
| MRI | 170 | Yes | $10^{3}$ to $10^{-5}$ | Yes | μg–mg | No | Yes |
| PET | 1000 | Yes | $10^{-11}$ to $10^{-12}$ | No | ng | No | Yes |
| SPECT (determined by collimator and head orientation) | 150–2000 | Yes | $10^{-10}$ to $10^{-11}$ | No | ng | No | Yes |
| Ultrasound | 30 | Limited (small volume) | High (not well characterized) | Yes | μg–mg | Yes | Yes |
| Photo-acoustic | 44–75 | (Small volume) | $10^{-7}$ | | | | Yes |
| Bioluminescence | >1000 | Limited (semi-quantitative) | $10^{-15}$ to $10^{-17}$ | Yes | μg–mg | Yes (?) | Yes (2D) |
| 2D fluorescence | >1000 | No | $10^{-9}$ to $10^{-12}$ | Yes | μg–mg | Yes | Yes |
| Fluorescence tomography | <1000 | Semi-quantitative | $10^{-9}$ to $10^{-12}$ | Yes | μg–mg | Yes | No |

utilization of various molecular imaging agents to probe specific pathways (Table 5.1), it has become necessary to incorporate numerous modalities to investigate the micro- and macro-biologic system for drug development.

## 5.2 Development and Validation of Model Platforms

To understand a drugs interaction within a biologic system, pertinent animal models have been established, from the simple subcutaneous (sc) injection of human cells to orthotopic models to genetically modified mouse models (GEMMs) incorporating knock-in, knockout, and CRISPR [14–16] technologies. Development of mouse model platforms for oncology drug development must also be validated utilizing histotechnological processes with respect to pathological standards such as angiogenesis (thymidine) and apoptosis (caspases, FITC-labeled Annexin V) [17]. Research groups such as the Biological Testing Branch, Division of Cancer Treatment and Diagnosis, and the Center for Advanced Preclinical Research (CAPR), Center for Cancer Research, both within NCI, have developed and fully characterized various xenograft models incorporating both human cells and tumor fragments and GEMM models (pancreas, lung, and ovarian) for testing new drugs against standard clinical therapies [18, 19].

In addition to the primary tumor, model platforms for the study of metastasis should be included in the repertoire [20]. Metastasis of cancer cells from a primary tumor is a leading cause of death [21], and early detection for timely therapeutic intervention with serial in vivo imaging would greatly improve clinical outcomes. Bioluminescence imaging (BLI) has been shown to be sensitive with the ability to image few cells, correlates to tumor volume as validated by gadolinium contrast MRI, and provides rapid imaging for high throughput [22]. Unfortunately, it only provides 2D images, and the depth penetration of light is limited to a few cm. On the other hand, utilizing BLI to first determine the presence of a metastatic signal (pre-screening technique), due to the BLI higher sensitivity and the higher throughput with respect to 3D small-bore modalities, BLI can improve utilization of higher-cost 3D modalities. This demonstrates that multimodality imaging does not mandate concurrent image acquisitions. Utilizing one modality to screen for presence of metastasis can greatly enhance utilization of another modality. Furthermore, due to the metastasis textural characteristics (i.e., echogenicity), a higher spatial resolution scanner (i.e., 30 μm for ultrasound) might not detect the metastatic lesion, whereas a lower spatial resolution scanner (i.e., 170 μm for MRI) can provide a higher-contrast signal (Fig. 5.1), such as in a T2* MRI sequence.

Further advancements in oncology animal models capturing the cell-autonomous (genetic and epi-genetic) and non-cell autonomous (stromal) aspects of tumor heterogeneity improve the understanding of patient-specific responses to therapy (precision medicine); thus cancer researchers have instituted patient-derived xenograft (PDX) animal model studies [23, 24]. The heterogeneity in the tumor fragment can be attributed to the tumor matrix that is transplanted with the tumor. This matrix tends to persist as the tumor grows, eventually becoming permeated and dissolving into the tumor. While there is mild heterogeneity seen in cellular base xenografts,

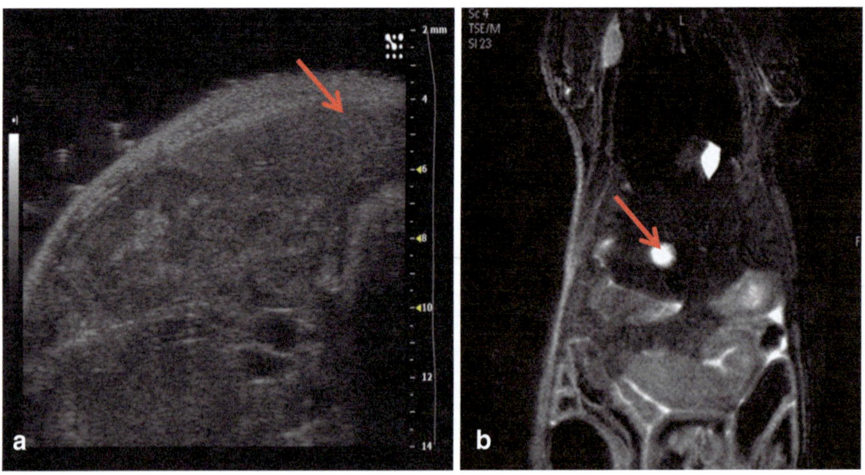


**Fig. 5.1** These two images demonstrate the marked difference in image contrast (red arrow) for a metastatic lesion (**a**) isoechoic with capsule in a B-mode ultrasound scanner (30 μm image resolution) and (**b**) high-contrast T2 signal in a 3 T MRI (150 μm image resolution) image with coils specific for small animals. While the resolution of US is greater than MRI, it is more difficult to compare lesions longitudinally on US. However, small animal US provides greater capability for dynamic characterization of individual lesions

the heterogeneity in the fragment group is far more typical of what is seen throughout the tumor growth in this group and is hypothesized to be more representative of the native in vivo microenvironment.

## 5.3 In Vivo Imaging for Oncology Translational Research

Within the vast array of technologies for small animal imaging, there are many opportunities to design imaging-based experiments to answer complex biological questions. However, in drug/therapy development, the requirements become more demanding and less flexible, shifting from an imaging experiment to an in vivo assay. To institute an in vivo assay, three key elements must exist: (1) relevant biomarker matched to imaging capability, (2) highly reproducible results, and (3) efficient throughput. Another aspect of importance is the ability to translate nonclinical assays to clinical research as needed for future development.

As previously discussed, this requires rigorous SOPs, a validated animal model, imaging equipment quality control, close attention to animal handling, quality control of probes, contrast agents, and radiopharmaceuticals, close adherence to acquisition protocols and a standardized analysis. The routine incorporation of both positive and negative pathological standards is also critical.

This conversion of the imaging experiment to an in vivo assay requires the development of an imaging assay platform that incorporates the appropriate validated model and a highly controlled imaging protocol designed to optimally measure the biomarker of interest. For the assay to be practical, both logistical and physiologic barriers need to be minimized by incorporation of such practices as keeping intravenous administrations to a minimum, reducing anesthesia sessions, and using systems that allow concurrent imaging of animals but allowing constant monitoring.

The implementation of small animal imaging in oncology drug development will now be described using three case study examples.

### 5.3.1 Case Studies

#### 5.3.1.1 Development of an Imaging Platform Consisting of an Animal Model and MR Technique

Colorectal cancer (CRC) is the third most common cancer in the United States, and the American Cancer Society estimated that in 2018 there will be 97,220 new cases and 50,630 deaths [25] and that chronic inflammation, such as ulcerative colitis and Crohn's disease, is associated with increased risk of CRC. To study colorectal cancer, animal models as an imaging assay platform can assist in the assessment of the initial stages of cancer and therapy response.

Conventional micro-endoscopes can provide appropriate information on colorectal cancer, i.e., imaging polyps, but risk perforating the colon or obstruction of the image due to bleeding associated with colitis. 3D in vivo imaging (i.e., virtual colonoscopy) can image the early phases of cancer, and utilizing various imaging agents can probe molecular pathways for characterization, including when altered by a drug. Clinically, X-ray CT has been a standard in performing virtual colonoscopy due to its rapid image acquisition. Preclinical X-ray CT scanners can provide information on polyp growth [26, 27] but are unable to provide the high tissue contrast necessary to discern normal and inflamed lumen tissue from surrounding tissue without resulting in high-radiation doses, especially in preclinical studies when serial imaging is required. Another method used in the clinic, magnetic resonance colonography (MRC), can discern normal and inflammatory tissue utilizing the dark lumen technique with either water or gas to expand the colon followed by IV contrast (Gd-chelate)-enhanced T1w MRI sequence, where water, if used as an enema, remains dark. T2w MRI is required to discern inflammatory tissue and also provides a rapid acquisition for high-throughput; unfortunately the water enema results in a bright signal. Preclinical MRI colorectal studies have utilized other techniques such as fecal tagging [28], and water enema [29], which unfortunately creates a bright lumen in T2w images making it difficult to distinguish normal from inflamed colonic mucosa, especially for researching drugs for chronic inflammation (pre-CRC).

Thus, the aim was to develop an imaging platform consisting of a mouse model and an adjusted MR protocol to study colorectal cancer. The in vivo MRI protocol should provide several important components: noninvasive serial imaging to monitor tumor progression and tissue inflammation, artifact-free imaging on T1w and T2w MRI sequences, enhanced image contrast (high signal-to-noise ratio: SNR) to discern inflamed and normal lumen tissue from surrounding tissue, quantitative imaging, and high throughput.

FVB/N mice were dosed (10 mg/kg, IP route) with a chemical carcinogen azoxymethane (AOM) and exposed 1 week later to the colonic irritant dextran sodium (DSS) [1–5% DSS dissolved in drinking water] for five cycles (5 days DSS and 16 days normal water) to develop the inflammation-induced colorectal cancer mouse model [30]. The virtual MRI colonoscopy technique implemented an enema procedure [31] using 1 mL of Fluorinert FC-770 (perfluorotri-n-butylamine, molecular formula of $C_{12}F_{27}N$), which does not create a MRI signal due to the lack of hydrogen atoms, and has been used in MRI imaging of human prostate cancer. Prior to MRI imaging, the enema was administered with a 20-G gavage syringe containing 0.6 mL of Fluorinert. The enema tubing was connected to a syringe pump and maintained at a continuous rate of 25 μL/min to maintain the colon distended during the imaging procedure. MRI (T1w and T2w) images were acquired pre- and post-contrast (gadopentetate dimeglumine (Gd-DTPA), 0.2 mmol/kg, IV injection, 150 μL/min infusion). Figures 5.2 and 5.3 demonstrate the utilization of a Fluorinert enema with MRI for virtual colonoscopy in the development of an animal model assay platform for the study of inflamed CRC. This standardized technique, incorporating both modality and animal model, provides for an assay platform for the study of drug efficacy studies for chronic inflammation (pre-CRC). In addition, the ability to fuse images from other modalities to virtual MRC can further enhance the information on molecular pathways in drug efficacy studies.

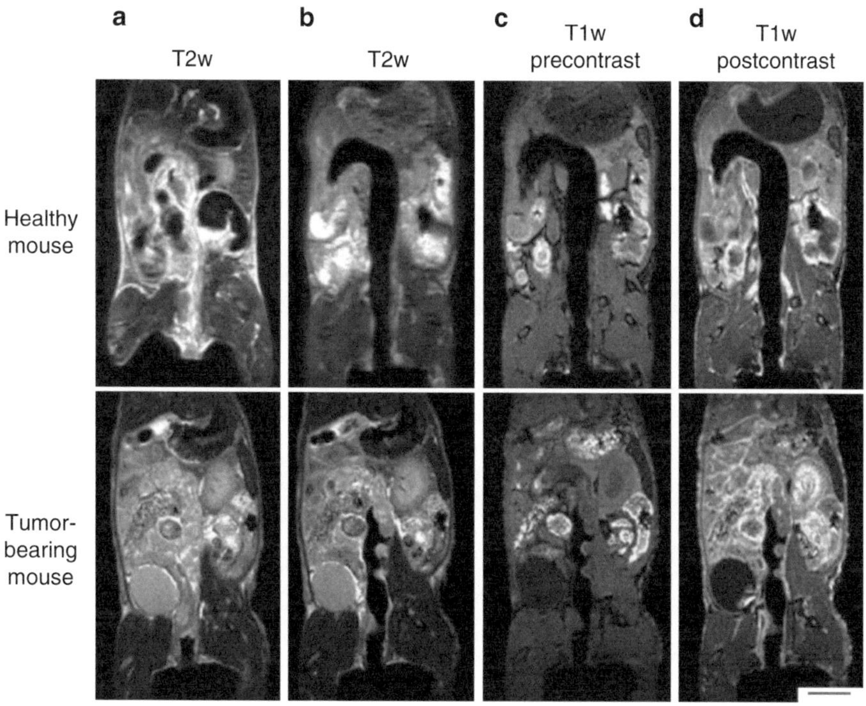


**Fig. 5.2** 3.0 T MR images of the mouse colon of healthy (top) and tumor-bearing mice (bottom row). (**a**) Coronal T2w image before Fluorinert enema infusion. (**b–d**) After Fluorinert enema: T2w image (**b**); T1w pre-contrast (**c**); and T1w post-contrast (Gd-DTPA) image (**d**). Scale bar, 5 mm. Ileva L et al. Nature Protocols, (2014), 9(11), 178–2682–2692. DOI:https://doi.org/10.1038/nprot.2014.178

### 5.3.1.2 Probe Validation: Nuclear Versus Optical Imaging

One major aspect for the development of an oncology drug as an in vivo assay is the development of a relevant biomarker matched to an imaging capability. One such drug, panitumumab (Vectibix), an anti-HER1 mAb, is a fully human mAb with minimal immunogenicity when injected intravenously. It is FDA approved for the treatment of HER1-expressing colorectal cancers, and is being evaluated in patients with other types of HER1-expressing cancers, such as breast, lung, head and neck, renal, and ovarian tumors [32]. The epidermal growth factor receptor (EGFR, erb1, HER1) is a glycoprotein belonging to subclass I of the tyrosine kinase receptor super family [33], disregulated in a variety of cancers [34], and is associated with disease progression and treatment resistance. Panitumumab binds to the domain III of HER1 and is rapidly internalized, leading to downregulation of cell surface HER1. It also arrests the cell cycle and inhibits tumor growth by suppressing the production of proangiogenic factors (VEGF, IL-8) by tumor cells [35].

To investigate a labeled panitumumab to risk-stratify clinical patients or as an intraoperative diagnostic probe for image-guided surgery, various subcutaneous

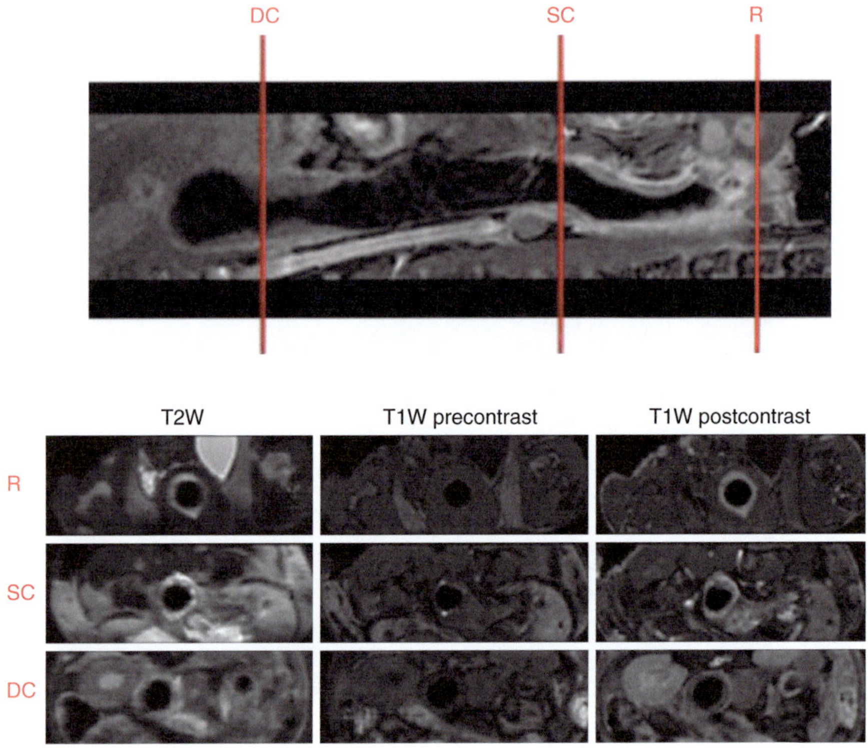


**Fig. 5.3** The MRI sagittal colon plane (top) provides the relative positions in the transverse plane for the rectum (R), sigmoid colon (SC) and descending colon (DC). The relative planes in the transverse slices (bottom) are generated from the T1w and T2w coronal 3D images demonstrating the Fluorinert enema and the enhancement of the lumen for studying inflammation in pre-CRC drug studies. Ileva L et al. Nature Protocols, (2014), 9(11), 178–2682–2692. DOI:https://doi.org/10.1038/nprot.2014.178

athymic nude female breast cancer xenograft tumor cell models have been developed with respect to HER1 expression (MDA-MB-469; high HER1, MDA-MB-231; mid HER1, and BT-474; low HER1). Panitumumab can be dual-labeled with a fluorescence dye (i.e., IRDye 800, optical component) and a radionuclide (i.e., 111-indium, SPECT component and/or 89-zirconium, PET component) for testing the various scenarios and models to visualize and quantify panitumumab uptake into the tumor and organs. Dual labeling is more time-consuming and costly compared to single labeling, with the caveat that quantified results of a single-labeled drug can be compared (molecular probe uptake into the tumor and organs) between modalities while utilizing standard animal handling techniques. Furthermore, dual labeling of radionuclides (i.e., SPECT and PET) might not be feasible due to the higher-energy PET photons (511 keV) penetration of the lower photon energy SPECT collimators. Figure 5.4 demonstrates dual-modality PET/CT coronal slices for the biodistribution of [$^{89}$Zr] panitumumab in various tumor-bearing animal

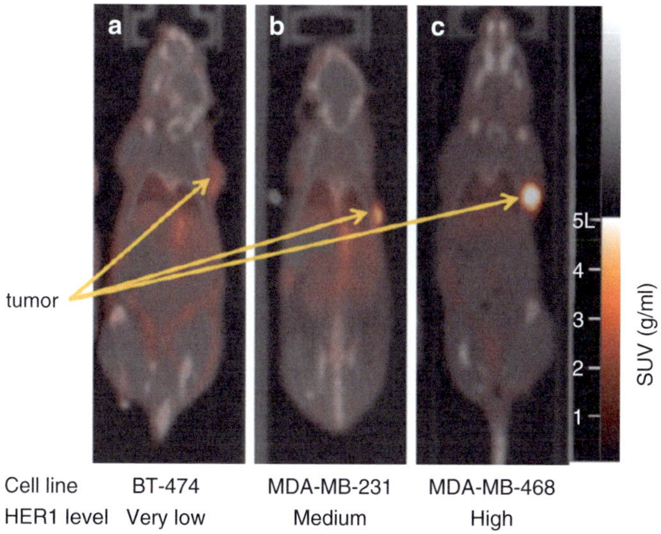


**Fig. 5.4** Coronal slices demonstrating tumor uptake of $^{89}$Zr-panitumumab in various subcutaneous athymic nude female xenograft models; 10.18 ± 1.24 MBq of $^{89}$Zr-panitumumab were administered intravenously via tail vein, and a 5-min CT scan followed by a 30-min static PET scan was performed at 96 h postinjection. The probe uptake into the tumor correlates with the HER1 protein expression. Orange arrows point to the representative HER1 tumors. Sibaprasad Bhattacharyya et al. Zirconium-89 labeled panitumumab: a potential immuno-PET probe for HER1-expressing carcinomas. Nuclear Medicine and Biology, 2013, 40, 451–457, https://doi.org/10.1016/j.nucmedbio.2013.01.007

models and correlation to the HER1 protein expression [36]. The same HER1 expression animal model(s) were later implemented in an epi-fluorescence imaging study for the determination of [IRDye 800]-labeled panitumumab for an intraoperative diagnostic probe for image-guided surgery [37]. Figure 5.5 demonstrates epi-fluorescence imaging for the HER1 animal models, resulting in similar probe uptake into the tumor(s) with respect to the HER1 expression, and Fig. 5.6 exhibits the correlation of the panitumumab imaging probe ([$^{89}$Zr] and [IRDye 800]) uptake between the different modalities. This case study demonstrates that providing a modulated signal (i.e., HER1) will enable the quantitative investigation of the underlining biomarker and that quantitation of the molecular biomarker with respect to the modality (i.e., fluorescence or nuclear probe) provides for an accurate technique for comparison between modalities without the necessity of costly dual labeling.

#### 5.3.1.3 Multimodality Probe Development

Inhalation of asbestos fibers is the primary cause of malignant pleural mesothelium (MPM), a highly lethal cancer affecting the lung pleural, and has been shown to result in increased tissue HER1 expression [38]. X-ray CT and MRI have difficulties distinguishing nonmalignant features of scarring and fibrosis from tumor tissue;

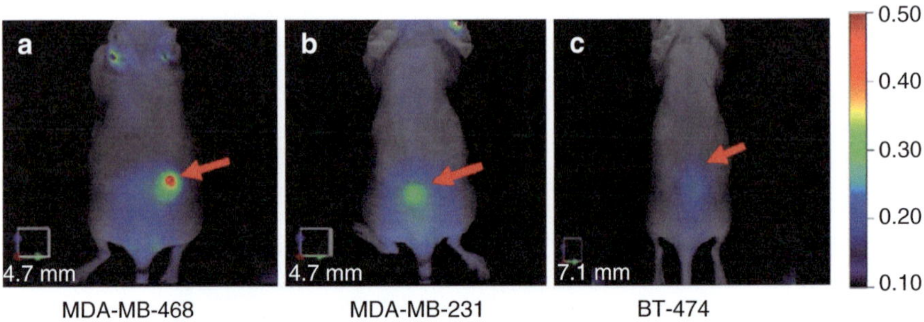


**Fig. 5.5** 2D epi-fluorescence images of panitumumab–IRDye800 of various HER1-expressing tumor (MDA-MB-468, MDA-MB-231, BT474) bearing athymic nude female mouse models at 24 h postinjection (100 mL of 1 mg mL$^{-1}$ conjugate). Red arrows point to the representative HER1 tumors. The probe uptake into the tumor correlates with the HER1 protein expression. Trace amounts of tracer accumulated in ears probably due to the inflammation caused by ear punch. Sibaprasad Bhattacharyya et al. Synthesis and biological evaluation of panitumumab–IRDye800 conjugate as a fluorescence imaging probe for EGFR-expressing cancers. Med. Chem. Commum., 2014, DOI: https://doi.org/10.1039/c4md00116h

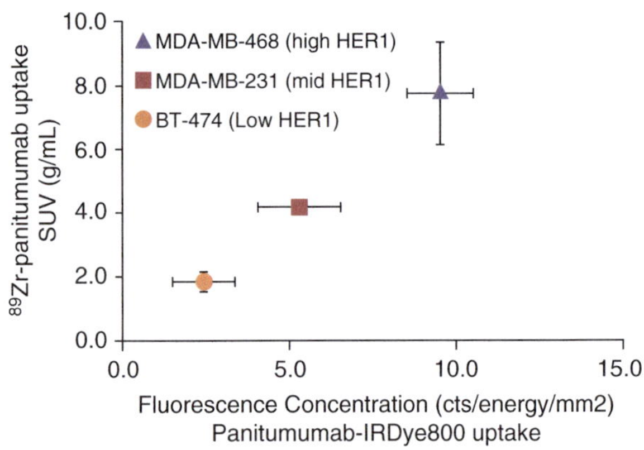


**Fig. 5.6** Uptake of panitumumab labeled with IRDye800 or [$^{89}$Zr] in different tumor xenografts with high, medium, and low EGFR expression, as measured by radioactive counts or fluorescence, is highly correlated. Sibaprasad Bhattacharyya et al. Synthesis and biological evaluation of panitumumab–IRDye800 conjugate as a fluorescence imaging probe for EGFR-expressing cancers. Med. Chem. Commum., 2014, DOI: https://doi.org/10.1039/c4md00116h

while the high specificity of [$^{18}$F]FDG PET for tumor imaging is successful, unfortunately high FDG uptake is also observed in benign inflammatory processes. Unfortunately, the X-ray CT used in most preclinical scanners is designed for PET photon attenuation correction and/or anatomical-functional image fusion that does not provide for the ability to segment the various tissues (organs). Therefore, localization of the molecular probe in the tumor is poorly differentiated from the surrounding tissues, which can result in significant quantitative issues depending on the preclinical animal model. Nyak et al. [39] studied MPM in an orthotopic (NCI-H226 and MSTO-211H mesothelium cells) MPM mouse model, fusing

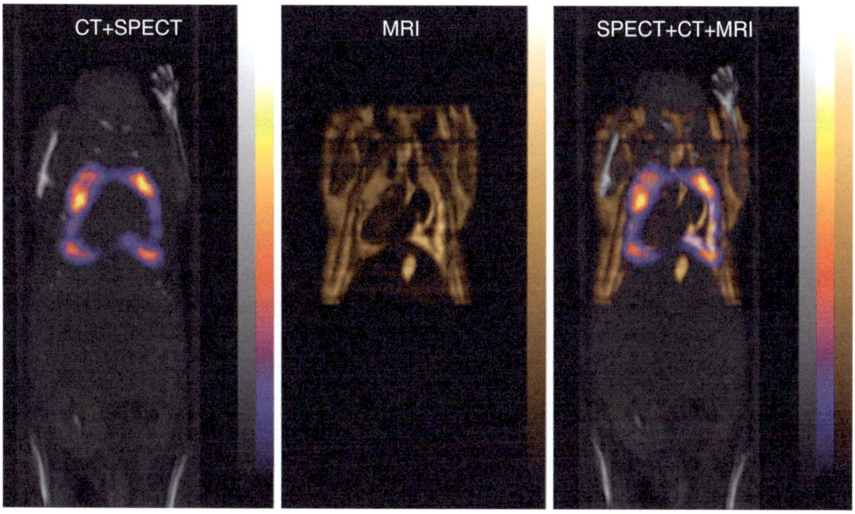


**Fig. 5.7** Representative coronal sections in female athymic (NCr) nu/nu mouse bearing orthotopic NCI-H226 cells injected intravenously via tail vein with 2.0 MBq of $^{111}$In-CHX-A"-DTPA–panitumumab. Images were acquired 5 days after the injection of radiolabeled panitumumab. Radiology 267, 2013: 173–182. DOI: https://doi.org/10.1148/radiol.12121021

panitumumab labeled with [$^{111}$In] for SPECT/CT with anatomic images from a 3.0 T MRI, exhibited in Fig. 5.7, demonstrating the enhancement of a multimodality study for both diagnostic and prognostic tools for the enhancement in the classification and assessment of the patients' disease state.

#### 5.3.1.4 Case Studies Summary

The first examples (colorectal cancer and panitumumab) demonstrate the validation strategy that coupling an imaging probe or technique with an imaging platform(s) has the characteristics of an assay. Specifically, in the panitumumab case, providing a modulated signal over the relevant biological scale will enable the quantitative investigation of the underlining biomarker, which in this case is HER1. The last case (multimodality probe development) demonstrates the enhancement of multimodalities to improve tissue characterization by reducing the false-positive features of a single modality. This case study further exemplifies in a multimodality study that establishing SOPs for equipment QC, animal handling, anesthesia, and image quantitation that utilizes animal models that are validated with respect to pathological standards is essential for the development of in vivo imaging assays.

#### 5.3.1.5 Future

Multimodality preclinical imaging will be a requirement in oncology drug development to understand the effect of treatment(s) on the various molecular pathways transforming the tumor microenvironment. The incorporation of patient-derived xenografts (PDX) into preclinical and co-clinical studies will correspondingly require multimodalities due to the multi-scale in tumor heterogeneity.

To understand and analyze these spatially distinct regions within a tumor as probed by multimodalities, investigators are incorporating texture analysis (fractals and lacunarity) [40–42]. For example, Dominietto et al. evaluated pattern analysis in a murine efficacy study investigating tumor angiogenesis [43]. MRI studies evaluated tumor blood volume and permeability at baseline and post-therapy. The authors evaluated the MRI images utilizing both standard histogram analyses (average values within a region of interest (ROI)) and pattern analysis (shape and texture) and concluded that the standard histogram was insensitive to determine therapeutic response, while pattern analysis appeared to be sensitive to tumor textural changes due to treatment. In addition to evaluating texture analysis in different modalities for clinical and preclinical studies, other authors are evaluating the effect of reconstruction algorithms on textural analysis, such as in PET imaging due to the limited number of projections and higher noise component [44, 45].

To improve the integration of multimodalities, especially for co-clinical endeavors, the National Cancer Informatics Program, which includes several NIH/National Cancer Institute Divisions (Center for Biomedical Informatics and Information Technology (CBIIT)), Cancer Imaging Program/Division of Cancer Treatment and Diagnosis), and several academic and industry partners, implemented a working group and developed a standard radiological image header (DICOM) for small animal imaging (Work Group 30: http://dicom.nema.org/dicom/geninfo/Strategy.pdf). These standards provide a framework for quantitative comparison of multimodality preclinical and clinical image sets utilizing identical image analysis algorithms. These headers will also allow for the incorporation and fusion of ex vivo pathological slides and molecular analysis, to include the full spectrum and multi-scale aspects for understanding the tumor microenvironment.

The future of multimodality imaging in small animals will provide an important basis for in vivo assay development in oncology drug discovery, improving quantitative measurement of therapeutic response, understanding the tumor microenvironment, and the integration of emerging fields such as radiomics and radiogenomics for improved patient outcomes.

**Acknowledgments** This project has been funded in whole or in part with federal funds from the National Cancer Institute, National Institutes of Health, under Contract No. HHSN261200800001E. The content of this publication does not necessarily reflect the views or policies of the Department of Health and Human Services, nor does mention of trade names, commercial products, or organizations imply endorsement by the U.S. Government Modified per federal agency (NIH).

Frederick National Laboratory for Cancer Research is accredited by AAALAC International and follows the Public Health Service Policy for the Care and Use of Laboratory Animals. Animal care was provided in accordance with the procedures outlined in the "Guide for Care and Use of Laboratory Animals" (National Research Council, 2011; National Academies Press, Washington, D.C.).

## References

1. DiMasi JA, Grabowski HG, Hansen RW. Innovation in the pharmaceutical industry: new estimates of R&D costs. J Health Econ. 2016;47:20–33. https://doi.org/10.1016/j.jhealeco.2016.01.012.

2. Thomas DW, Burns J, Audette J, Carroll A, et al. Clinical development success rates 2006–2015, BioIndustry analysis. http://www.amplion.com/clinical-development-success-rates?hsCtaTracking=7e38cfe3-248d-440b-a7e4-c038acfa6eb2%7Ca6180579-5624-4deb-ac76-35b512407bd1
3. Vanhove C, Bankstahl JP, Krämer SD, Visser E, Belcari N, Vandenberghe S. Accurate molecular imaging of small animals taking into account animal models, handling, anesthesia, quality control and imaging system performance. EJNMMI Phys. 2015;2:31. https://doi.org/10.1186/s40658-015-0135-y.
4. Kinahan P, Fletcher JW. PET/CT standardized uptake values (SUVs) in clinical practice and assessing response to therapy. Semin Ultrasound CT MR. 2010;31(6):496–505. https://doi.org/10.1053/j.sult.2010.10.001.
5. Sha W, Ye H, Iwamoto KS, Wong K-P, Wilks MQ, Stout D, McBride W, Huang S-C. Factors affecting tumor $^{18}$F-FDG uptake in longitudinal mouse PET studies. EJNMMI Res. 2013;3:51. https://doi.org/10.1186/2191-219X-3-51.
6. Adiseshaiah PP, Patel NL, Ileva LV, Kalen JD, Haines DC, McNeil SE. Longitudinal imaging of cancer cell metastasis in two preclinical models: a correlation of noninvasive imaging to histopathology. Int J Molecul Imaging. 2014;2014:102702. https://doi.org/10.1155/2014/102702.
7. Fuchs K, Kukuk D, Mahling M, Quintanilla-Martinez L, Reischl G, Reutershan J, Lang F, Rocken M, Pichler BJ, Kneilling M. Impact of anesthetics on 3′-[$^{18}$F]fluoro-3′-deoxythymidine ([$^{18}$F]FLT) uptake in animal models of cancer and inflammation. Mol Imaging. 2013:1–11. https://doi.org/10.2310/7290.2012.00042.
8. Maier FC, Kneilling M, Reischl G, Cay F, Bukala D, Schmid A, Judenhofer MS, Röcken M, Machulla H-J, Pichler BJ. Significant impact of different oxygen breathing conditions on noninvasive in vivo tumor-hypoxia imaging using [$^{18}$F]-fluoro-azomycinarabino- furanoside ([$^{18}$F] FAZA). Radiat Oncol. 2011;6:165. https://doi.org/10.1186/1748-717X-6-165.
9. Fueger BJ, Czernin J, Hildebrandt I, Tran C, Halpern BS, Stout D, Phelps ME, Weber WA. Impact of animal handling on the results of $^{18}$F-FDG PET studies in mice. J Nucl Med. 2006;47(6):999–1006.
10. Fuchs K, Kukuk D, Reischl G, Foller M, Eichner M, Reutershan J, Lang F, Rocken M, Pichler BJ, Kneilling M. Oxygen breathing affects 3′-deoxy-3'-$^{18}$F-fluorothymidine uptake in mouse models of arthritis and cancer. J Nucl Med. 2012;53:823–30. https://doi.org/10.2967/jnumed.111.101808.
11. Hildebrandt IJ, Helen S, Weber WA. Anesthesia and other considerations for in vivo imaging of small animals. ILAR. 2008;49(1):17–26. https://doi.org/10.1093/ilar.49.1.17.
12. Ileva LV, Bernardo M, Patel NL, Riffle LA, Graff-Cherry C, Robinson C, Difilippantonio S, Kalen JD. Challenges in performing preclinical imaging in a large cohort therapeutic efficacy study of murine cancer models. 64th AALAS National Meeting, Baltimore, MD, October 29, 2013.
13. Honndorf VS, Schmidt H, Wehrl HF, Wiehr S, Ehrlichmann W, Quintanilla-Martinez L, Barjat H, Ricketts S-A, Pichler BJ. Quantitative correlation at the molecular level of tumor response to docetaxel by multimodal diffusion-weighted magnetic resonance imaging and [$^{18}$F]FDG/[$^{18}$F]FLT positron emission tomography. Mol Imaging. 2014;(1) https://doi.org/10.2310/7290.2014.00045.
14. Yang H, Wang H, Shivalila CS, Cheng AW, Shi L, Jaenisch R. One-step generation of mice carrying reporter and conditional alleles by CRISPR/cas-mediated genome engineering. Cell. 2013;154(6):1370–9. https://doi.org/10.1016/2013.08.022.
15. The Jackson Laboratory, Bar Harbor, ME USA, https://www.jax.org/.
16. Tentler JJ, Tan AC, Weekes CD, Jimeno A, Leong S, Pitts TM, Arcaroli JJ, Messersmith WA, Gail Eckhardt S. Patient-derived tumor xenografts as models for oncology drug development. Nat Rev Clin Oncol. 2012;9:338–50. https://doi.org/10.1038/nrclinonc.2012.61.
17. Elmore S. Apoptosis: a review of programmed cell death. Toxicol Pathol. 2007;35(4):495–516.
18. Biological Testing Branch, Division of Cancer Diagnostics and Treatment, NCI, NIH: https://dtp.cancer.gov/organization/btb/default.htm
19. Center for Advanced Preclinical Research, Center for Cancer Research, NCI, NIH: https://ccr.cancer.gov/capr

20. van Marion DMS, et al. Studying cancer metastasis: Existing models, challenges and future perspectives. Crit Rev Oncol Hematol. 2015;97:107–17. https://doi.org/10.1016/j.critrevonc.2015.08.00.
21. Chaffer CL, Weinberg RA. A perspective on cancer cell metastasis. Science. 2011;331(6024):1559–64. https://doi.org/10.1126/science.1203543.
22. Troy T, Jekic-McMullen D, Sambucetti L, Rice B. Quantitative comparison of the sensitivity of detection of fluorescent and bioluminescence reporters in animal models. Mol Imaging. 2004;3(1):9–23.
23. Siolas D, Honnon GJ. Patient-derived tumor xenografts: transforming clinical samples into mouse models. Cancer Res. 2013;73(17):5315–9. https://doi.org/10.1158/0008-5472.CAN-13-1069.
24. Cassidy JW, Caldas C, Bruna A. Maintaining tumor heterogeneity in patient-derived tumor xenografts. Cancer Res. 2015;75(15):2963–8. https://doi.org/10.1158/0008-5472.CAN-15-0727.
25. American Cancer Society. Colorectal cancer. 2018.
26. Durkee BY, Weichert JP, Halberg RB. Small animal micro-CT colonography. Methods. 2010;50:36–41. https://doi.org/10.1016/j.ymeth.2009.07.008.
27. Boll H, Bag S, Nölte IS, Wilhelm T, Kramer M, Groden C, Böcker U, Brockmann MA. Double-contrast micro-CT colonoscopy in live mice. Int J Color Dis. 2011;26:721–7. https://doi.org/10.1007/s00384-011-1181-0.
28. Larsson AE, et al. Magnetic resonance imaging of experimental mouse colitis and association with inflammatory activity. Inflamm Bowel Dis. 2006;12:478–85.
29. Herborn CU, et al. Dark lumen magnetic resonance colonography in a rodent polyp model: initial experience and demonstration of feasibility. Investig Radiol. 2004;39:723–7.
30. Ileva LV, Bernardo M, Young MR, Riffle LA, Tatum JL, Kalen JD, Choyke PL. In vivo MRI virtual colonography in a mouse model of colon cancer. Nat Protoc. 2014;9(11):2682–92. https://doi.org/10.1038/nprot.2014.178.
31. Young MR, Ileva LV, Bernardo M, Riffle LA, Jones YL, Kim YS, Colburn NH, Choyke PL. Monitoring of tumor promotion and progression in a mouse model of inflammation-induced colon cancer with magnetic resonance colonography. Neoplasia. 2009;11(3):237–46. https://doi.org/10.1593/neo.81326.
32. Wu M, Rivkin A, Pham T. Panitumumab: human monoclonal antibody against the epidermal growth factor receptors for the treatment of metastatic colorectal cancer. Clin Ther. 2008;30:14–30. https://doi.org/10.1016/j.clinthera.2008.01.014.
33. Burgess AW. EGFR family: structure physiology signaling and therapeutic targets. Growth Factors. 2008;26:263–74. https://doi.org/10.1080/08977190802312844.
34. Ciardiello F, Tortora G. Anti-epidermal growth factor receptor drugs in cancer therapy. Expert Opin Investig Drugs. 2002;11:755–68. https://doi.org/10.1517/13543784.11.6.755.
35. Yang XD, Xia XC, Corvalan JR, Wang P, Davis CG. Development of ABX-EGF, a fully human anti-EGF receptor monoclonal antibody, for cancer therapy. Crit Rev Oncol Hematol. 2001;38:17–23. https://doi.org/10.1016/S1040-8428(00)00134-7.
36. Bhattacharyya S, Kurdziel K, Wei L, Riffle L, Kaur G, Hill GC, Jacobs PM, Tatum JL, Dorosho JH, Kalen JD. Zirconium-89 labeled panitumumab: a potential immuno-PET probe for HER1-expressing carcinomas. Nucl Med Biol. 2013;40:451–7. https://doi.org/10.1016/j.nucmedbio.2013.01.007.
37. Bhattacharyya S, Patel NL, Wei L, Riffle LA, Kalen JD, Hill GC, Jacobs PM, Zinn KR, Rosenthal E. Synthesis and biological evaluation of panitumumab–IRDye800 conjugate as a fluorescence imaging probe for EGFR-expressing cancers. Med Chem Commum. 2014; https://doi.org/10.1039/c4md00116h.
38. Faux SP, Houghton CE, Hubbard A, Pat- rick G. Increased expression of epidermal growth factor receptor in rat pleural mesothelial cells correlates with carcinogenicity of mineral fibres. Carcinogenesis. 2000;21(12):2275–80. https://doi.org/10.1093/carcin/21.12.2275.
39. Nayak TK, Bernardo M, Milenic DE, Choyke PL, Brechbiel MW. Orthotopic Pleural Mesothelioma in Mice: SPECT/CT and MRI Imaging with HER1-and HER2-targeted

Radiolabeled Antibodies. Radiology. 2013;267:173–82. https://doi.org/10.1148/radiol.12121021.

40. Asselin M-C, O'Connor JPB, Boellaard R, Thacker NA, Jackson A. Quantifying heterogeneity in human tumours using MRI and PET. Eur J Cancer. 2012;48:447–55. https://doi.org/10.1016/j.ejca.2011.12.025.
41. Soares F, Janela F, Pereira M, Seabra J, Freire MM. 3D lacunarity in multifractal analysis of breast tumor lesions in dynamic contrast-enhanced magnetic resonance imaging. IEEE Trans Image Process. 2013;22(11):4422–35. https://doi.org/10.1109/TIP.2013.2273669.
42. Goh V, Sanghera B, Wellsted DM, Sundin J, Halligan S. Assessment of the spatial pattern of colorectal tumor perfusion estimated at perfusion CT using two-dimensional fractal analysis. Eur Radiol. 2009;19:1358–65. https://doi.org/10.1007/s00330-009-1304-y.
43. Dominietto M, Lehmann S, Keist R, Rudin M. Pattern analysis accounts for heterogeneity observed in MRI studies of tumor angiogenesis. Magn Reson Med. 2013;70:1481–90. https://doi.org/10.1002/mrm.24590.
44. Leijenaar RTH, Nalbantov G, Carvalho S, van Elmpt WJC, Troost EGC, Boellaard R, Aerts HJWL, Gillies RJ, Lambin P. The efect of SUV discretization in quantitative FDG-PET Radiomics: the need for standardized methodology in tumor texture analysis. Sci Rep. 2015;5:11075. https://doi.org/10.1038/srep11075.
45. Buvat I, Orlhac F, Soussan M. J Nucl Med. 2015;56(11):1642–4. https://doi.org/10.2967/jnumed.115.163469.

# Investigation of Transporter-Mediated Drug-Drug Interactions Using PET/MRI

6

Thomas Wanek, Alexander Traxl, Claudia Kuntner-Hannes, and Oliver Langer

## 6.1 Introduction

Drug disposition consists of absorption, distribution, metabolism and excretion (ADME). For these processes, drugs and drug metabolites have to cross cellular membranes in different organs and tissues (e.g. intestine, liver, kidney, etc.). In many cases, movement of drugs and drug metabolites across cellular membranes is not just a passive diffusion-mediated process but occurs via saturable transmembrane transporters belonging either to the solute carrier (SLC) or adenosine triphosphate-binding cassette (ABC) families. When two drugs, which are recognized by the same SLC or ABC transporters, are co-administered, one drug may induce transporter inhibition/saturation and thereby change the disposition of the other drug as compared to when the drugs are administered alone. This phenomenon has been termed transporter-mediated drug-drug interaction (DDI), which is of great concern in drug development as this may impact drug safety and efficacy [1]. In many cases, changes in the activity of transmembrane transporters may lead to

T. Wanek (✉)
AIT Austrian Institute of Technology GmbH, Center for Health & Bioresources, Biomedical Systems, Seibersdorf, Austria

BioImaging Austria (CMI), Seibersdorf, Austria
e-mail: thomas.wanek@ait.ac.at

A. Traxl · C. Kuntner-Hannes
AIT Austrian Institute of Technology GmbH, Center for Health & Bioresources, Biomedical Systems, Seibersdorf, Austria

O. Langer
AIT Austrian Institute of Technology GmbH, Center for Health & Bioresources, Biomedical Systems, Seibersdorf, Austria

Departments of Clinical Pharmacology and Biomedical Imaging and Image-Guided Therapy, Medical University of Vienna, Vienna, Austria



C. Kuntner-Hannes, Y. Haemisch (eds.), *Image Fusion in Preclinical Applications*, https://doi.org/10.1007/978-3-030-02973-9_6

pronounced changes in drug tissue distribution (e.g. liver, kidneys, brain) without changes in drug plasma pharmacokinetics [2]. To assess such tissue DDIs in vivo, a methodology is needed to measure drug tissue concentration levels. The non-invasive nuclear imaging method positron emission tomography (PET) is a very powerful tool for measuring tissue distribution of drugs radiolabelled with positron-emitting radionuclides, such as carbon-11 ($^{11}C$, half-life: 20.4 min) or fluorine-18 ($^{18}F$, half-life: 109.8 min), in animals or humans [3, 4]. Dedicated small-animal PET systems allow to measure rodents (mice, rats) with high sensitivity and good spatial resolution. As PET does not provide anatomical information, it needs to be combined with anatomical imaging to allow for better definition of organs or tissues of interest. In this chapter, we provide a case report of how PET in combination with magnetic resonance imaging (MRI) can be used to assess transporter-mediated DDIs in different organs of the mouse.

## 6.2 Sequential PET/MR Imaging

PET is a nuclear imaging technique, which allows the non-invasive determination of the concentration of a radiolabelled drug in tissues. PET utilizes the radioactive decay of positron-emitting radionuclides such as $^{18}F$ or $^{11}C$ to track the localization of a radiolabelled drug in the body. Besides its non-invasive nature, the advantages of PET are its high sensitivity (only a few μg of the radiolabelled compound is required) and its high temporal resolution enabling the detection of concentration changes in tissues within several seconds. However, PET has a limited spatial resolution (~2 mm) and provides only limited anatomical information, which may make it difficult to allocate the PET signal to designated organs or tissues. Contrary to PET, MRI delivers anatomical information with high spatial resolution and excellent soft-tissue contrast. Especially the organs involved in elimination processes of drugs, such as the liver, gall bladder and kidneys, can be clearly visualized by MRI without the use of additional contrast agents. Therefore, MRI adds truly complementary information to PET data by facilitating the definition of these organs when used in a combined way.

Combining PET and MRI can be achieved by either software fusion or hardware combination. Several combined or simultaneous preclinical PET/MRI systems are now commercially available [5, 6]. PET/MRI images can also be achieved by scanning the subjects on two stand-alone scanners in a serial manner and subsequent image fusion. This is usually performed using software tools that load the image data of both modalities and either use an automated or manually adjusted rigid or non-rigid transformation matrix. Image fusion is simplified when fiducial markers or anatomical landmarks are identified. In contrast to clinical imaging, where the patient has to be moved between the two scanners, often resulting in different patient positions on the bed, this is not an issue in preclinical imaging. Here, not the animal alone but the whole imaging chamber (including the retained and anaesthetized animal) can be moved from one scanner to the other. Thus, subject position usually remains the same within the chamber, which simplifies image fusion as only rotation, and translation can occur. Especially for benchtop MRI systems with low-field strength and permanent magnets, this combination is ideal, as they can be installed almost anywhere and at relatively low costs and due to the low-field strength do not interfere with photomultiplier tube (PMT)-based PET scanners. Placing them near dedicated small-animal PET scanners, these benchtop MRI systems enable sequential PET/MRI acquisitions, combining the advantages of both modalities and thus providing results comparable to a full-scale sequential PET and a high-field-strength MR measurement, with few restrictions in terms of MR image quality and animal handling. Another advantage of using two separate scanners is that both scanners can be utilized alone, especially for studies when no multimodal image information is required, i.e. only PET or only MRI scans. Moreover, the maintenance costs and technical challenges are lower for stand-alone systems compared to combined scanners.

A prerequisite for sequential PET/MR imaging is a multimodality imaging chamber, which provides heating, inhalation anaesthesia and animal monitoring (i.e. temperature, respiratory and heart rate). Moving the animal from one bed to the next results in changes in animal position, which complicates image fusion. To overcome this issue, one can either use a multimodality imaging chamber with proper connectors on each modality (also providing anaesthesia gas support, heating and animal monitoring) or use the imaging chamber from one modality in all other modalities, by means of adapters. Benchtop MRI systems often have an animal bed where the RF coil is directly mounted over the animal, and the combination is slid into the scanner. Thus, the construction of an adapter and mounting plate for the PET scanner is necessary.

Another premise for sequential PET/MR imaging is the correct and precise co-registration of the two data sets. The used methods and algorithms have to be practical but also robust so that they can be used on a daily basis. One of the main challenges is that for certain radiotracers PET images show no or little anatomical information. As a result, it may be difficult to co-register them with MR images just based on morphological structures, which can be seen in both images. Furthermore, new radiotracers for PET examinations become more and more specific in terms of their interaction with the molecular imaging target, so that they show little retention in nontarget tissues (e.g. radiolabelled antibodies). As mentioned before, apart from anatomical landmarks, fiducial markers can be used to simplify image fusion. Fiducial markers are small objects that are visible in the used modality, in this case PET and MRI. Small glass capillaries (inner diameters of 0.1–0.5 mm) are typically used that can be filled with radioactive solution and sealed with adhesive. At least three markers are then attached to or inside the animal chamber next to the animal, in such a way that they are visible in the final images of all modalities (means for MRI that they have to be inside the RF coil) but do not interfere with the imaging study. The drawback of using these markers is that they increase the time for animal positioning and, even if they are small, may complicate positioning of the animals inside the imaging chamber. Especially for the animal holder used in the benchtop 1T MRI system, space is very limited as the whole body RF coil is positioned very close over the animal bed and animal, already restricting the size of the scanned animal. Another method to generate a transformation matrix for image fusion is phantom measurements using a multimodality phantom that incorporates landmarks (e.g. intersection points) that are visible in all modalities. When imaging the phantom with the identical protocol as used for the animal studies (same acquisition and reconstruction protocol), it is possible to generate a rigid transformation matrix that can afterwards be applied to the animal data.

## 6.3 Role of Transporters in Drug Disposition

Transport of drugs can be mediated by members of two large families of transmembrane transporter proteins, the SLC and the ABC transporter family. Both transporter families use cellular energy, either in form of electrochemical gradients or ATP hydrolysis to transport compounds against concentration gradients. These pumps can move substrates in (influx) or out (efflux) of cells. In humans and rodents, ABC and SLC transporters are for instance expressed in the liver, intestine, blood-brain barrier (BBB), blood-testis barrier, placenta and kidney. For instance, in the liver organic anion-transporting polypeptides (e.g. OATP1B1/SLCO1B1, OATP1B3/SLCO1B3) in the basolateral membrane of hepatocytes mediate the uptake of drugs from blood in the liver, whereas canalicular ABC transporters (e.g. multidrug resistance-associated protein 2, MRP2/ABCC2 or breast cancer resistance protein, BCRP/ABCG2) can promote the excretion of drugs and drug metabolites from hepatocytes into bile [7]. Impaired function of basolateral uptake transporters due to DDIs, disease or genetic polymorphisms can lead to marked increases in blood concentrations of drugs, which can cause severe side effects (e.g. rhabdomyolysis for statin drugs). Impaired function of canalicular efflux transporters, on the other hand, can lead to drug accumulation in the liver, which may cause hepatotoxicity.

Erlotinib is a reversible tyrosine kinase inhibitor (TKI) of the epidermal growth factor receptor (EGFR) that has been approved for treatment of advanced, metastatic non-small cell lung cancer (NSCLC) and advanced, unresectable or metastatic pancreatic cancer. Approximately 10% of NSCLC patients in the Western population harbour an activating mutation in their EGFR genes (e.g. the exon 19 deletion delE746-A750 or the exon 21 point mutation L858R) resulting in higher response rates to treatment with erlotinib or gefitinib, another EGFR-inhibiting TKI, as compared with patients with wild-type EGFR [8].

Erlotinib undergoes extensive metabolism in humans and is mainly excreted via the hepatobiliary pathway leading to high concentration levels in the liver and bile [9]. Erlotinib is a substrate of BCRP, P-glycoprotein (Pgp/ABCB1), organic anion transporter 3 (OAT3/SLC22A8) and organic cation transporter 2 (OCT2/SLC22A2). Efflux transport by P-gp and BCRP at the BBB results in a low extent of brain distribution of erlotinib [10–12]. Less information is currently available about how far drug transporters influence distribution of erlotinib to other organs than the brain and to what extent transporters are involved in erlotinib excretion.

In this study, we assessed the influence of transporters on $^{11}$C-erlotinib organ distribution and excretion. To accomplish this, we performed sequential MRI/PET measurements with a microdose of $^{11}$C-erlotinib (<5 μg) without and with the co-injection of a pharmacologic dose of unlabelled erlotinib or pretreatment with elacridar, a dual P-gp/BCRP inhibitor, in wild-type mice. In addition, we scanned P-gp/Bcrp knockout (Abcb1a/b$^{(-/-)}$Abcg2$^{(-/-)}$) mice. The recorded images were co-registered, and organ concentration levels were further quantitated in order to obtain pharmacokinetic data. The technical challenges of this project lie in the use of two different scanner systems and the development of devices and procedures which allow a rapid and secure exchange of the animals from one scanner setup to the other including anaesthesia, animal heating and monitoring systems. In order to obtain images suitable for image co-registration, animals have to be moved from one scanner to the other in a way that no body movement of the animal occurs. In addition, for optimal image co-registration, a transformation matrix to correct for the position mismatch of the reconstructed MRI and PET images has to be calculated.

## 6.4 Material and Methods

### 6.4.1 Phantom Study

For calculation of a transformation matrix for image co-registration, a multimodality phantom was constructed. This phantom consists of a polypropylene cylinder filled with a mixture of $H_2O$, $CuSO_4$ (1 g/L) and NaCl (3.6 g/L) in which a flexible tube containing $^{18}$F-fluoride solution was inserted. The tube was positioned inside the phantom in such a way that intersection points at different planes were created. These intersection points were used as markers to define the position of the phantom in the MRI or PET scanner. In total, the measurement phantom contained 15 intersection points. The last one was located 42 mm away from the first intersection point. Every kind of intersection point appeared three times. The phantom was positioned on the MRI animal bed with the whole body mouse coil, and images were acquired with the PET and the MR using the same acquisition and reconstruction protocols as used for the in vivo study. To co-register the phantom PET and MR images, the data set alignment wizard in AMIDE [13] was used, which allows the alignment of one image data set with another. For alignment, the rigid body registration method using fiducial markers was selected. Before the data alignment wizard could be used, the fiducial markers were defined on the PET and MR images at the positions of the intersection points. Transverse, coronal and sagittal views of the data set in AMIDE were used to identify the correct position of the fiducial markers as good as possible. For every fiducial marker, the centre location was documented before and after the PET, and MRI data sets were aligned. Afterwards the alignment was calculated by AMIDE using the Procrustes rigid body alignment algorithm without scaling. Position mismatch was calculated by alignment of the intersection points of the flexible tubes within the phantom in the MR and PET images resulting in a rigid transformation matrix that can be applied to all data sets. Here, the PET data set was defined as the fixed and the MR data set as the moving one, because the moving data set is interpolated after the transformations were applied. This should be avoided for the PET data set, as it is used for further analysis of the drug distribution data and interpolation could lead to falsified results. Evaluation of image co-registration accuracy was performed by calculation of the fiducial registration error (FRE) and the target registration error (TRE).

### 6.4.2 Animals

Female FVB and Abcb1a/b$^{(-/-)}$Abcg2$^{(-/-)}$ mice were obtained from Charles River and Taconic. At the time of experiment, animals were 10–15 weeks old and weighed 25.7 ± 2.7 g. Animals were housed in groups in individual ventilated cages (IVCs) under controlled environmental conditions. Animals were allowed to acclimatize >1 week prior to the experiments. The study was approved by the local Animal Welfare Committee in accordance with the Austrian Animal Experiments Act. All efforts were made to minimize the number of animals as well as pain or discomfort.

### 6.4.3 Anaesthesia and Monitoring

Anaesthesia was initiated using 2.5–3% isoflurane in oxygen and maintained at 1–1.5% for the imaging procedure. Body temperature and respiratory rate were constantly monitored during the whole experiment using an MR compatible small-animal monitoring and gating system. Animals were positioned prone on the mouse bed, front teeth were positioned inside the tooth bar and body temperature was maintained by warm water heating.

### 6.4.4 Experimental Design

Wild-type animals were divided into three groups: In the first group ($n$ = 4), a dynamic $^{11}$C-erlotinib PET scan was performed without any pretreatment. The second group of mice ($n$ = 6) underwent dynamic $^{11}$C-erlotinib PET scans at 20 min after intravenous pretreatment with the dual P-gp/BCRP inhibitor elacridar (10 mg/kg). In the third group, mice ($n$ = 4) underwent dynamic $^{11}$C-erlotinib PET scans, in which a pharmacologic dose of unlabelled erlotinib (10 mg/kg) was co-injected with $^{11}$C-erlotinib. In addition, Abcb1a/b$^{(-/-)}$Abcg2$^{(-/-)}$ mice ($n$ = 4) underwent a dynamic $^{11}$C-erlotinib PET scan without any pretreatment.

### 6.4.5 MRI

Anatomic MR imaging was performed using a benchtop MRI scanner (ICON 1 T, Bruker BioSpin GmbH, Ettlingen, Germany). The 1T MRI is equipped with a 1T permanent magnet with negligible fringe field. Its gradient coil provides 450 mT/m. The animal bed is equipped with a nose-cone for inhalation anaesthesia, fixation system (tooth bar), heating system (built-in water tubing in the animal bed), temperature and respiration monitoring. After positioning the animal on the animal bed and connection of all cables, the mouse whole body RF coil (length 80 mm) was slid over the animal and animal bed. Images were acquired using a modified three-dimensional T1-weighted gradient echo sequence (T1-fast low-angle shot) with the following parameters: echo time = 5 ms; repetition time = 25 ms; flip angle = 25°; field of view = 76 × 28 × 24 mm; matrix = 253 × 93; 32 slices; slice thickness = 0.75 mm; and scan time = 6.25 min.

### 6.4.6 PET

After MRI, the animal holder was transferred into the gantry of a Siemens microPET Focus220 system and secured using a custom-made mounting plate. A 10-min transmission scan using a $^{57}$Co point source was followed by a 90-min emission scan which was started at time of intravenous injection of $^{11}$C-erlotinib (26 ± 9 MBq, 1 ± 1 nmol, 0.10 mL, $n = 18$). The scanner acquires the data in list mode format with a 6 ns coincidence timing window and an energy window of 250–750 keV. The reported spatial image resolution was 1.8 mm FWHM within 1 cm radial offset and the detection sensitivity was 3% [14]. The PET emission data were sorted into 25 frames, which incrementally increased in time length from 5 s to 20 min. Images were reconstructed using Fourier rebinning followed by two-dimensional filtered backprojection. The reconstructed PET images consist of a 128 × 128 × 95 matrix with a final voxel size of 0.4 × 0.4 × 0.8 mm$^3$. The standard data correction protocol (normalization, decay correction, injection decay correction and attenuation correction) was applied to all data sets.

### 6.4.7 Image Analysis

Images were co-registered using the medical data examiner software AMIDE [13]. After applying the transformation matrix calculated from the phantom study to the animal data, volumes of interest (VOIs) were manually drawn over the whole brain, right lung, left ventricle of the heart, left kidney, liver, gallbladder, intestine and urinary bladder in the fused MR/PET images, and time-activity concentration curves expressed in standardized uptake values (SUV) from 0 to 90 min after injection of the radiolabelled compound were derived. It was assumed that the sum of radioactivity in the gallbladder and the intestine represented radioactivity in the bile excreted from the liver. From the time-activity curves, the area under the curve from time 0 to 90 min (AUC) was calculated using Prism 5.0 software (GraphPad Software).

Furthermore, a graphical analysis approach (integration plot) was used to estimate the rate constants for cerebral, hepatic, renal and pulmonary uptake ($k_{\text{uptake, brain}}$,

$k_{\text{uptake, liver}}$, $k_{\text{uptake, kidney}}$ and $k_{\text{uptake, lung}}$) and the rate constants for biliary and urinary excretion ($k_{\text{bile}}$, $k_{\text{urine}}$) of $^{11}$C-erlotinib. Rate constants for uptake were measured from 0.3 to 3.5 min after tracer injection for the brain, liver and kidney and from 0.6 to 4.5 min after tracer injection for the lung using the integration plot method [15] and the following equation:

$$\frac{X_{t,\text{brain}}}{C_{t,\text{blood}}} = \text{CL}_{\text{uptake,brain}} \times \frac{\text{AUC}_{0-t}}{C_{t,\text{blood}}} + V_{E,\text{brain}} \quad \frac{C_{t,\text{organ}}}{C_{t,\text{blood}}} = k_{\text{uptake}} \times \frac{\text{AUC}_{0-t,\text{blood}}}{C_{t,\text{blood}}} + V_E$$

where $C_{t,\text{ organ}}$ is the radioactivity concentration in the brain, liver, kidney or lung at time $t$ and $C_{t,\text{ blood}}$ is the radioactivity concentration in the left ventricle of the heart at time t. $\text{AUC}_{0\text{-}t,\text{ blood}}$ represents the area under the concentration-time curve in the left ventricle of the heart from time 0 to time $t$. $K_{\text{uptake}}$ can be obtained by performing linear regression analysis of a plot of $C_{t,\text{ organ}}/C_{t,\text{ blood}}$ versus $\text{AUC}_{0\text{-}t,\text{ blood}}/C_{t,\text{ blood}}$ and calculating the slope of the regression line. The unit of $k_{\text{uptake}}$ is mL blood per min per gram tissue (mL/min/g tissue), and it thus corresponds to $K_1$ from kinetic modelling of PET data [16]. $V_E$ is the y-intercept of the integration plot.

Rate constants for biliary and urinary excretion ($k_{\text{bile}}$ and $k_{\text{urine}}$) of $^{11}$C-erlotinib were measured from 8.8 to 65 min and 12.5 to 65 min after tracer injection, respectively, using the integration plot method and the following equations:

$$C_{t,\text{intestine}} = k_{\text{bile}} \times \text{AUC}_{0-t,\text{liver}} + V_E$$

and

$$C_{t,\text{urine}} = k_{\text{urine}} \times \text{AUC}_{0-t,\text{kidney}} + V_E$$

where $C_{t,\text{ intestine}}$ and $C_{t,\text{ urine}}$ are the radioactivity concentrations in the intestine (including the gall bladder and duodenum) and urine, respectively, at time t. $\text{AUC}_{0\text{-}t,\text{ liver}}$ and $\text{AUC}_{0\text{-}t,\text{ kidney}}$ represent the area under the concentration-time curve in the liver and the kidney, respectively, from time 0 to time $t$. $k_{\text{bile}}$ and $k_{\text{urine}}$ can be obtained by performing linear regression analysis of a plot of $C_{t,\text{ intestine}}$ versus $\text{AUC}_{0\text{-}t,\text{ liver}}$ and $C_{t,\text{ urine}}$ versus $\text{AUC}_{0\text{-}t,\text{ kidney}}$, respectively, and calculating the slope of the regression line. $V_E$ is the y-intercept of the integration plot.

## 6.5 Results and Discussion

Both employed imaging systems are located in the same room within close proximity (~3 m) enabling a quick transfer of animals from one scanner system to the other. The external stray field of the 1T MRI system is negligible and therefore no RF cage is needed. The integration of anaesthesia, animal warming and respiratory trigger in the MRI animal holder allows switching between both scanner systems without interruption of the entire animal handling systems. Hence, anaesthesia and warm water for temperature stabilization were constantly supplied, and animal vital parameters were monitored. All PET scans were performed after MRI with the mouse bed and mounted mouse whole body coil. This setup ensures that movement of the scanned subject is limited; however, the mounted coil increased attenuation and scatter for the PET measurements which has to be corrected by performing a transmission scan prior to the PET measurement. Including the attenuation information into PET image reconstruction leads to the same quantitative data as with PET-only scans [17].

With the aim to safely and reproducibly mount the MRI animal holder onto the PET system, a mounting plate was constructed that allows exact positioning of the completely assembled animal holder onto the PET system. This mounting plate consists of a bottom plate which exactly fits to the bed moving mechanism of the PET scanner, with a fixable clamp to secure the animal holder and a mounting guide to prevent any rotation of the animal holder and four adjustable screws (Fig. 6.1a). The mounting plate can be easily removed from the PET scanner if not needed. By

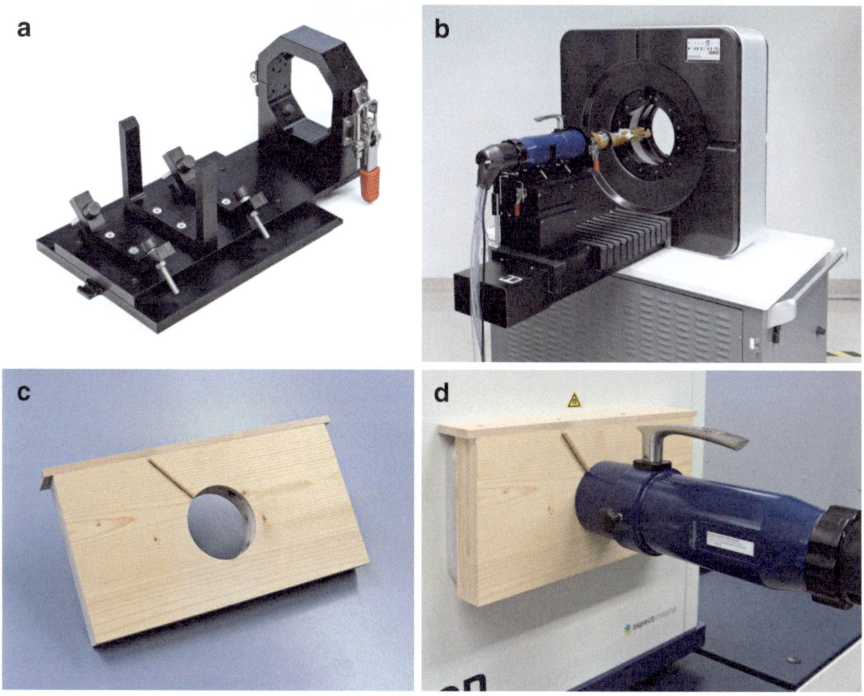

**Fig. 6.1** (**a**) Custom-made mounting plate for reproducible positioning of the Bruker animal holder on the PET system. (**b**) Bruker animal bed holder with mouse bed and mouse whole body coil mounted on a Siemens Focus220 microPET system. The animal bed holder can be moved from the ICON MRI to the PET system within seconds without interruption of anesthesia or animal warming. The employed system ensures that the animal position in the PET system corresponds exactly to the position in the previous MRI scan. (**c**) Home-made position finder for reproducible positioning of the Bruker animal holder in the MRI system. (**d**) Bruker animal bed holder with mouse bed and mouse whole body coil inserted into the ICON MRI

using this setup, the animal bed holder could be transferred from the MRI to the PET scanner within seconds without interruption of anaesthesia, animal warming or monitoring systems. Using the same horizontal and vertical positioning parameters of the bed moving mechanism of the PET scanner, a reproducible position inside the centre field of view could be achieved for all conducted experiments (Fig. 6.1b). To ensure a reproducible position of the animal bed inside the MRI scanner, a position finder was designed that can be easily attached to the Bruker ICON (Fig. 6.1c, d). All MRI and attenuation measurements could be performed within the 20-min pre-treatment phase with elacridar prior to injection of $^{11}$C-erlotinib.

Figure 6.2a shows the mouse whole body PET/MRI phantom that was constructed to calculate a transformation matrix for image co-registration of corresponding MRI and PET images. All 15 aligned intersections were clearly visible in

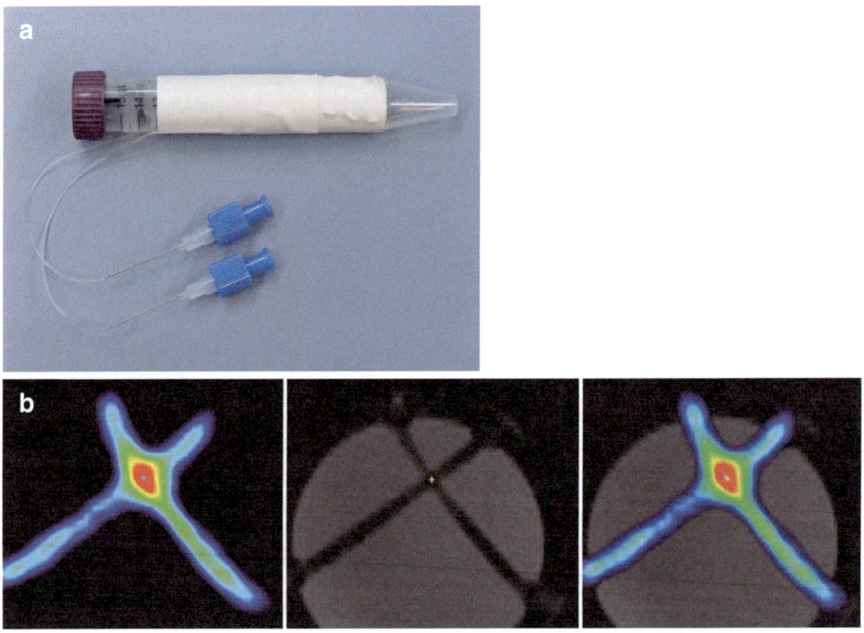


**Fig. 6.2** (**a**) Mouse whole body phantom used for the calculation of a transformation matrix to correct for images mismatch in fused MR and PET images. The phantom consists of a 15 mL polypropylene falcon tube filled with a solution of $CuSO_4$ and NaCl in water. In this falcon tube holes were drilled and a thin, flexible tube (inner diameter 0.8 mm) was threaded in such a way that it forms a series of intersections in all anatomical orientations. Prior to measurements the flexible tube was filled with a solution containing $^{18}$F-fluoride and the end of the flexible tube was secured with screw caps. (**b**) Transversal PET (left) and T1-weighted gradient echo MRI (middle) views of the mouse whole body phantom acquired using the Bruker animal holder equipped with the mouse whole body coil. The flexible tube filled with $^{18}$F-fluoride solution can be clearly distinguished from the background of the MRI phantom in both imaging modalities. A transformation matrix was calculated by placing fiducial markers at the 5 intersections visible in the PET and MR images and further rotation and transformation of one dataset to exactly match the position of the fiducial markers (right)

all MRI and PET images, which allowed placement of sufficient fiducial markers from which the position mismatch between reconstructed MRI and PET images could be calculated (Fig. 6.2b). The final calculated mean registration error from the centre location between both reconstructed MRI and PET images was <0.2 mm ($FRE_{mean}$ = 0.18 ± 0.02 mm and $TRE_{mean}$ = 0.18 ± 0.02 mm). Deviations caused by rotation errors were almost negligible.

Quantitative data of organ concentration of $^{11}$C-erlotinib were derived after image co-registration of corresponding MRI and PET images. The transformation matrix calculated from the phantom studies could be used for all scanned animals without further modifications.

We assessed the effect of transporters on tissue distribution and excretion of $^{11}$C-erlotinib. In Fig. 6.3 representative co-registered PET/MR images and in Fig. 6.4

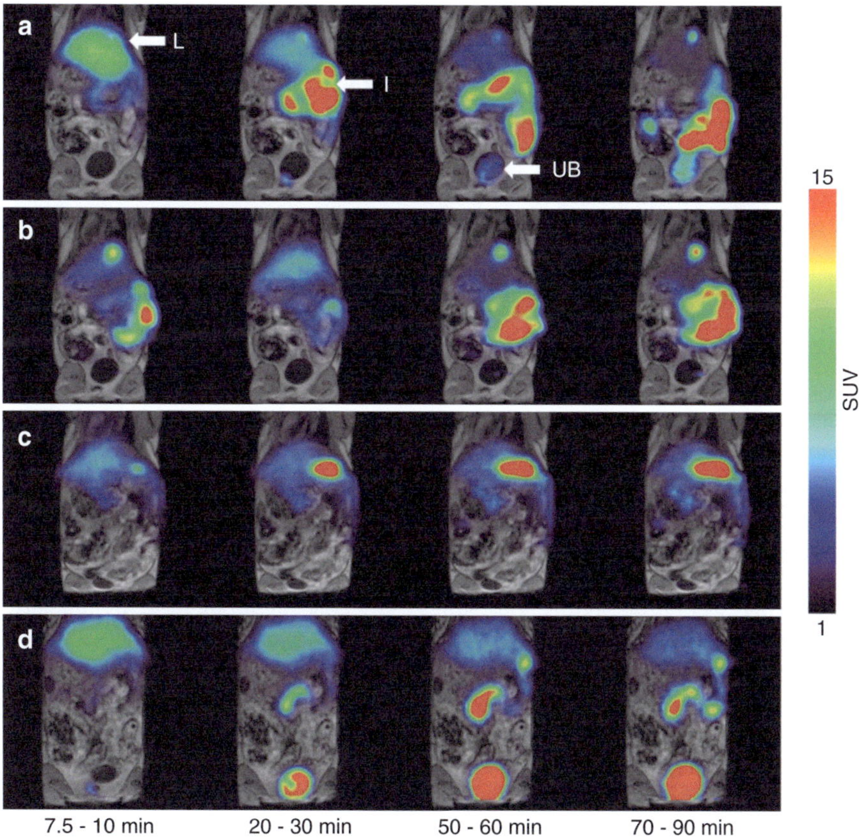


**Fig. 6.3** Serial coronal whole-body PET/MR images of $^{11}$C-erlotinib in a wild-type mouse (**a**), a wild-type mouse pretreated at 20 min before PET with 10 mg/kg elacridar (**b**), a wild-type mouse co-injected with a pharmacologic dose (10 mg/kg) of erlotinib (**c**), and an Abcb1a/b$^{(-/-)}$Abcg2$^{(-/-)}$ mouse. Anatomical structures are indicated by arrows: L = liver, I = intestine; UB = urinary bladder

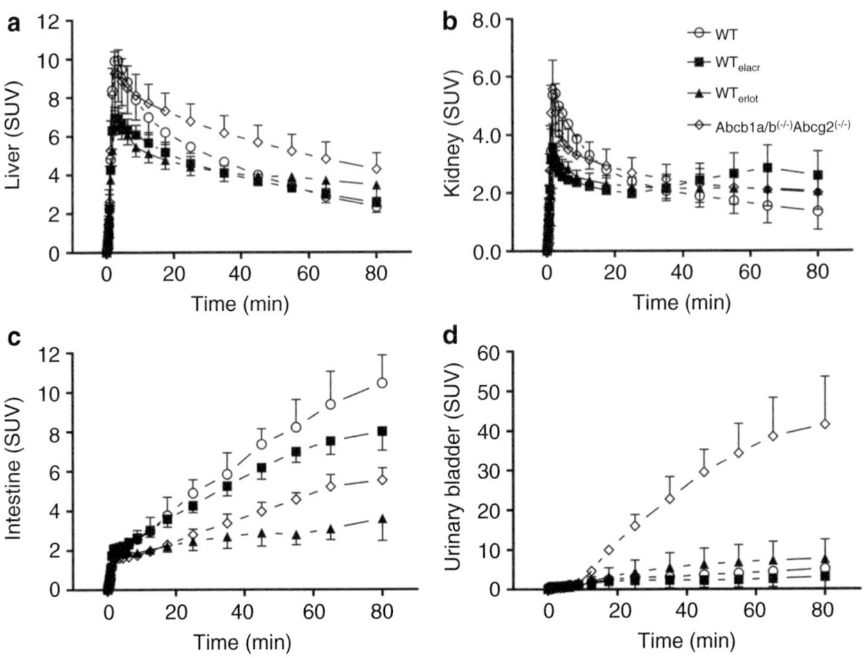


**Fig. 6.4** Time-activity curves (mean SUV ± SD) of [11]C-erlotinib in (**a**) liver, (**b**) kidney, (**c**) intestine and (**d**) urinary bladder of wild-type mice (WT), wild-type mice pretreated at 20 min before PET with 10 mg/kg elacridar ($WT_{elacr}$), wild-type mice co-injected with a pharmacologic dose (10 mg/kg) of erlotinib ($WT_{erlot}$), and Abcb1a/b$^{(-/-)}$Abcg2$^{(-/-)}$ mice

time-activity curves in different organs are shown. There was a clear visual difference in the biodistribution of [11]C-erlotinib between wild-type and Abcb1a/b$^{(-/-)}$Abcg2$^{(-/-)}$ mice, in that there was prolonged retention of radioactivity in the liver of Abcb1a/b$^{(-/-)}$Abcg2$^{(-/-)}$ mice with less radioactivity in the intestine and more radioactivity in the urinary bladder (Figs. 6.3 and 6.4). Also, in erlotinib co-injected wild-type mice, a prolonged liver retention and less radioactivity content in the intestine were observed. Blood radioactivity concentrations were significantly higher in erlotinib co-injected mice (AUC in left ventricle of the heart, wild-type microdose: 91.0 ± 2.4 SUV*min, pharmacologic dose: 135.8 ± 13.3 SUV*min). Integration plot analysis revealed significantly lower $k_{bile}$ values in Abcb1a/b$^{(-/-)}$Abcg2$^{(-/-)}$ mice and in erlotinib co-injected wild-type mice as compared with wild-type mice which received a microdose of [11]C-erlotinib (Fig. 6.5d). $k_{urine}$ values were significantly increased in Abcb1a/b$^{(-/-)}$Abcg2$^{(-/-)}$ mice (Fig. 6.5e). Taken together this suggests that Bcrp and P-gp at the canalicular membrane of hepatocytes mediated biliary excretion of [11]C-erlotinib and/or its radiolabelled metabolites. These transporters appear to become saturated when a pharmacologic dose of erlotinib was co-injected leading to a reduction in $k_{bile}$. Interestingly, the P-gp/BCRP inhibitor elacridar appeared to be not able to inhibit hepatic P-gp/Bcrp (Fig. 6.5d).

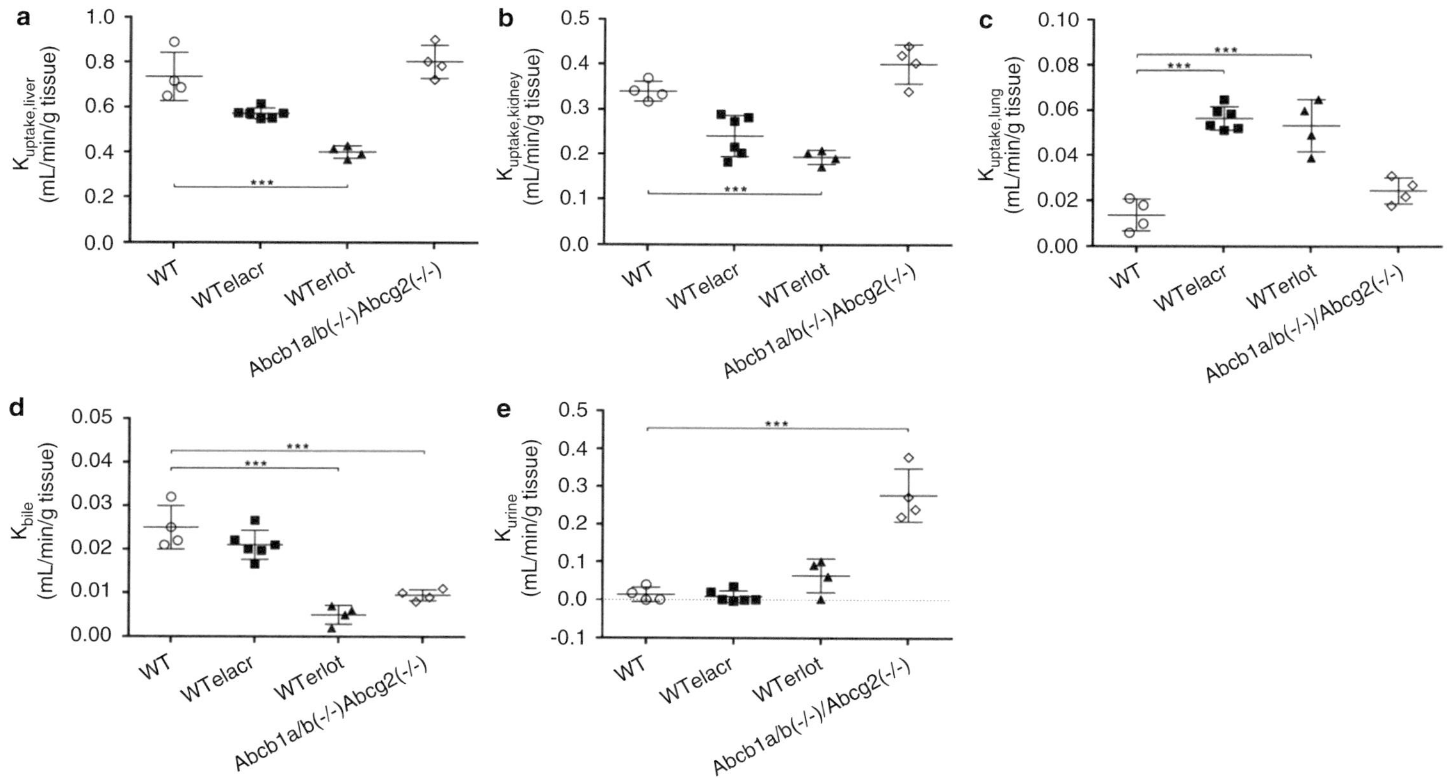


**Fig. 6.5** Rate constants for (**a**) hepatic, (**b**) renal, and (**c**) pulmonary uptake (mean $k_{uptake,organ}$ ± SD) as well as for (**d**) biliary and (**e**) urinary excretion (mean $k_{fluid}$ ± SD) of $^{11}$C-erlotinib in wild-type mice (WT), wild-type mice pretreated at 20 min before PET with 10 mg/kg elacridar ($WT_{elacr}$), wild-type mice co-injected with a pharmacologic dose (10 mg/kg) of erlotinib ($WT_{erlot}$), and Abcb1a/b$^{(-/-)}$Abcg2$^{(-/-)}$ mice. ***$P < 0.001$, 1-way ANOVA with Bonferroni's multiple comparison test

The increase in $k_{urine}$ values in Abcb1a/b$^{(-/-)}$Abcg2$^{(-/-)}$ mice was consistent with a shift from hepatobiliary to renal excretion in absence of canalicular Bcrp/P-gp activity (Figs. 6.4d and 6.5e).

In the brain, AUCs and $k_{uptake, brain}$ values were significantly higher in elacridar-treated wild-type mice as well as in Abcb1a/b$^{(-/-)}$Abcg2$^{(-/-)}$ mice (Fig. 6.6), which confirmed that P-gp and Bcrp restricted brain distribution of $^{11}$C-erlotinib and that elacridar can completely inhibit P-gp/Bcrp at the mouse BBB. Interestingly, also co-injection of unlabelled erlotinib increased $k_{uptake, brain}$, which indicated partial transporter saturation at the BBB and which suggests a potential utility of high-dose erlotinib to enhance brain uptake of other dual P-gp/BCRP substrates.

Integration plot analyses were conducted to obtain the rate constants for uptake of $^{11}$C-erlotinib from blood into the liver, kidney and lung (Fig. 6.5). In the liver and kidney, $k_{uptake}$ was significantly lower for the pharmacologic dose than for the microdose, consistent with saturation of basolateral uptake transporters in hepatocytes and kidney cells, such as OATPs or organic anion transporters. This is in line with previous findings that TKIs including erlotinib are competitive inhibitors of these SLC transporters [18]. This most likely also accounted for the significantly higher blood AUCs for the pharmacologic dose as compared with the microdose. In the lungs, $k_{uptake}$ was significantly increased in animals receiving the pharmacologic dose (Fig. 6.5a–c).

In other organs than the brain and the lungs, elacridar pretreatment exerted no significant effect on $^{11}$C-erlotinib distribution. In the liver, there was a trend for decreased $k_{uptake}$ values following elacridar, which indicates that elacridar may inhibit erlotinib uptake transporters in hepatocytes.

An advantage of performing transporter studies in mice is the fact that the whole animal is within the field of view of the PET scanner so that all mouse organs can

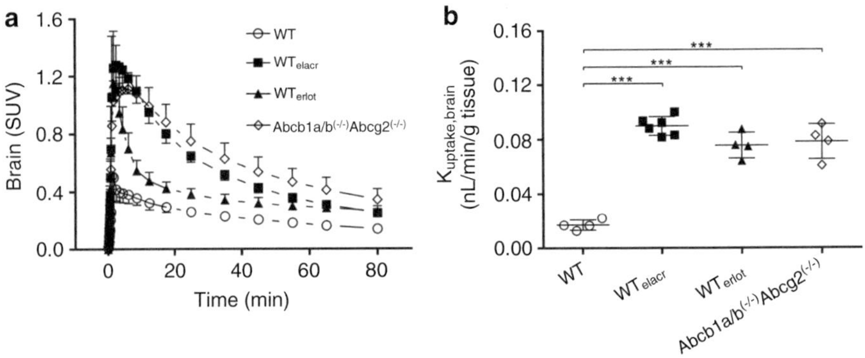


**Fig. 6.6** (**a**) Time-activity curves (mean SUV ± SD) and (**b**) rate constants for whole brain uptake (mean $k_{uptake,brain}$ ± SD) of $^{11}$C-erlotinib in wild-type mice (WT), wild-type mice pretreated at 20 min before PET with 10 mg/kg elacridar (WT$_{elacr}$), wild-type mice co-injected with a pharmacologic dose (10 mg/kg) of erlotinib (WT$_{erlot}$), and Abcb1a/b$^{(-/-)}$Abcg2$^{(-/-)}$ mice. ***$P < 0.001$, 1-way ANOVA with Bonferroni's multiple comparison test

be dynamically measured at the same time to assess transporter effects. This also allows obtaining an image-derived blood input function, e.g. from the left ventricle of the heart, which is needed for quantitative analysis of the PET data to obtain quantitative parameters of transporter activity (e.g. $k_{\mathrm{bile}}$, $k_{\mathrm{uptake}}$). In addition, the commercial availability of transgenic transporter knockout mice enables to elucidate the effect of complete absence of one or several transporters on drug disposition. The additional benefit of including the MRI measurements into this study was the easier delineation of the organs of interest to extract the time-activity curves from the PET image data. Especially the definition of the urinary bladder in the erlotinib co-injected group was challenging without MR images as there was neither radioactivity uptake in the bladder nor in adjacent regions. Using the MRI data for the definition of the VOIs leads to a more precise extraction of PET information resulting in reproducible quantitative data. Moreover, it also enables the user to create an organ atlas template based on the MR images that could be then used to facilitate analysis of PET-only image data.

## 6.6 Conclusion

The performed imaging study could confirm that Bcrp, P-gp and SLC uptake transporters (e.g. OATPs) influence in vivo disposition of $^{11}$C-erlotinib and thereby affect its distribution to normal and potentially also tumour tissue. Saturable transport of erlotinib leads to non-linear pharmacokinetics, which needs to be considered when attempting to predict the organ distribution of erlotinib in tumour patients using PET scans with a microdose of $^{11}$C-erlotinib. Moreover, erlotinib may be a perpetrator of transporter-mediated DDIs when being combined with other drugs that are transported by BCRP, P-gp and OATPs. This may for instance lead to changes in hepatic disposition of victim drugs. Inhibition of P-gp and BCRP at the BBB appears to be a promising approach to enhance brain distribution of erlotinib to increase its efficacy in the treatment of NSCLC brain metastases.

The motivation for combining PET with MR in this study was the interest in the in vivo whole body disposition of $^{11}$C-erlotinib. A special focus was on the extraction of uptake and excretion rate constants in multiple organs, which made it necessary to define the whole brain, lung, left ventricle of the heart, kidney, liver, gallbladder, intestine and urinary bladder on the PET images. Some of these organs (e.g. gallbladder) were only clearly visible in the MR images, and thus image data analysis without the corresponding individual anatomical image would have been impossible. As the desired quantitative PET parameters that need to be generated were clear from the beginning of the study, the inclusion of the MR was already fixed in the study planning phase.

Moreover, this study demonstrated that a 1T benchtop MRI system, which offers many features of high-field strength preclinical MRI at low running costs, enables high-quality PET/MRI studies when the MRI is installed in close proximity to the dedicated PET scanner. The time difference between the scans is negligible and animal monitoring and heating is constantly provided. The setup introduced in this study showed that combined PET and anatomical MR imaging is feasible and can be a valuable research tool in elucidating the role of transporters in drug disposition.

For future dual modality experiments, the following list of items needs to be considered in the study planning phase:

1. Selection of the animal model based on the research question
   (a) Definition of age and sex.
   (b) Inclusion of a control group.
   (c) Adequate sample size calculation.
2. Selection of the functional imaging modality based on the research question
   (a) PET versus SPECT.
   (b) Available radioisotopes and labelling procedures.
   (c) Biological versus physical half-life.
3. Selection of the anatomical imaging modality based on the research question
   (a) Definition of organs (regions) of interest—thereafter selection of modality (CT or MR) and selection of MR coils (head or whole body).
   (b) Needed resolution—definition of voxel sizes.
4. Definition if integrated or sequential dual modality imaging is needed.
5. Definition if individual (especially MR) imaging is needed.

## References


1. Nies AT, Schwab M, Keppler D. Interplay of conjugating enzymes with OATP uptake transporters and ABCC/MRP efflux pumps in the elimination of drugs. Expert Opin Drug Metab Toxicol. 2008;4(5):545–68.
2. Kusuhara H, Sugiyama Y. In vitro-in vivo extrapolation of transporter-mediated clearance in the liver and kidney. Drug Metab Pharmacokinet. 2009;24(1):37–52.
3. Langer O. Use of PET imaging to evaluate transporter-mediated drug-drug interactions. J Clin Pharmacol. 2016;56(Suppl 7):S143–56.
4. Wagner CC, Langer O. Approaches using molecular imaging technology -- use of PET in clinical microdose studies. Adv Drug Deliver Rev. 2011;63(7):539–46.
5. Nagy K, Toth M, Major P, Patay G, Egri G, Haggkvist J, et al. Performance evaluation of the small-animal nanoScan PET/MRI system. J Nucl Med. 2013;54(10):1825–32.
6. Wehrl HF, Amend M, Thielcke A. Multimodal imaging and image fusion. In: Kiessling F, Pichler BJ, Hauff P, editors. Small Animal imaging basics and practical guide. Berlin: Springer International Publishing; 2017. p. 491–507.
7. Kalliokoski A, Niemi M. Impact of OATP transporters on pharmacokinetics. Br J Pharm. 2009;158(3):693–705.
8. Hammerman PS, Janne PA, Johnson BE. Resistance to epidermal growth factor receptor tyrosine kinase inhibitors in non-small cell lung cancer. Clin Cancer Res. 2009;15(24):7502–9.
9. Ling J, Johnson KA, Miao Z, Rakhit A, Pantze MP, Hamilton M, et al. Metabolism and excretion of erlotinib, a small molecule inhibitor of epidermal growth factor receptor tyrosine kinase, in healthy male volunteers. Drug Metab Dispos. 2006;34(3):420–6.
10. Agarwal S, Manchanda P, Vogelbaum MA, Ohlfest JR, Elmquist WF. Function of the blood-brain barrier and restriction of drug delivery to invasive glioma cells: findings in an orthotopic rat xenograft model of glioma. Drug Metab Dispos. 2013;41(1):33–9.

11. de Vries NA, Buckle T, Zhao J, Beijnen JH, Schellens JH, van Tellingen O. Restricted brain penetration of the tyrosine kinase inhibitor erlotinib due to the drug transporters P-gp and BCRP. Investig New Drugs. 2012;30(2):443–9.
12. Kodaira H, Kusuhara H, Ushiki J, Fuse E, Sugiyama Y. Kinetic analysis of the cooperation of P-glycoprotein (P-gp/Abcb1) and breast cancer resistance protein (Bcrp/Abcg2) in limiting the brain and testis penetration of erlotinib, flavopiridol, and mitoxantrone. J Pharmacol Exp Ther. 2010;333(3):788–96.
13. Loening AM, Gambhir SS. AMIDE: a free software tool for multimodality medical image analysis. Mol Imaging. 2003;2(3):131–7.
14. Tai YC, Ruangma A, Rowland D, Siegel S, Newport DF, Chow PL, et al. Performance evaluation of the microPET focus: a third-generation microPET scanner dedicated to animal imaging. J Nucl Med. 2005;46(3):455–63.
15. Takashima T, Wu C, Takashima-Hirano M, Katayama Y, Wada Y, Suzuki M, et al. Evaluation of breast cancer resistance protein function in hepatobiliary and renal excretion using PET with 11C-SC-62807. J Nucl Med. 2013;54(2):267–76.
16. Takano A, Kusuhara H, Suhara T, Ieiri I, Morimoto T, Lee YJ, et al. Evaluation of in vivo P-glycoprotein function at the blood-brain barrier among MDR1 gene polymorphisms by using 11C-verapamil. J Nucl Med. 2006;47(9):1427–33.
17. Schmid A, Schmitz J, Mannheim JG, Maier FC, Fuchs K, Wehrl HF, et al. Feasibility of sequential PET/MRI using a state-of-the-art small animal PET and a 1 T benchtop MRI. Mol Imaging Biol. 2013;15(2):155–65.
18. Johnston RA, Rawling T, Chan T, Zhou F, Murray M. Selective inhibition of human solute carrier transporters by multikinase inhibitors. Drug Metab Dispos. 2014;42(11):1851–7.

# 7 Multimodality Preclinical Imaging in Inflammatory Diseases

Paul D. Acton

## 7.1 Inflammation and Its Role in Disease

Inflammation is a complex biological response to invasion of the host by pathological organisms and other harmful stimuli, such as bacterial infection, damaged cells, wounds, or irritants. The inflammatory response typically is the first in a cascade of defense mechanisms, and plays a vital role in the elimination of harmful pathogens, and in the initial repair of damaged tissue.

Acute inflammation is a normal response of the body to maintain homeostasis after injury or host invasion. The early inflammatory response is considered part of the innate immune system, since it is not specific to any particular pathogen [1]. The process typically occurs rapidly after the initial harmful stimulus and involves the local activation of a number of immune cells, particularly macrophages, dendritic cells, Kupffer cells, and mast cells. Depending on the stimulus, different cell types recognize different features of the stimulus or pathogen and release pro-inflammatory cytokines, such as interleukin-1 (IL-1) and tumor necrosis factor alpha (TNF$\alpha$), and chemokines, such as interleukin-8 (IL-8). The resulting vasodilation and increase in vascular permeability triggers accumulation of plasma proteins, such as fibrin and antibodies, at the site of injury. This leads to many of the classical features of inflammation, such as redness from the increase in blood flow to the tissue, and edema and swelling from the accumulation of fluid and proteins.

Leukocytes, including macrophages and neutrophils, extravasate into tissue, migrating to the site of injury along a concentration gradient of the locally released chemoattractants [2]. Depending on the nature of the tissue damage, phagocytosis of pathogens or cellular debris occurs. If the inflammatory response was triggered by a wound, aggregation of platelets and fibrin occurs, leading to coagulation and clot formation and eventually tissue repair.

P. D. Acton (✉)
Janssen Research and Development, Johnson & Johnson, Spring House, PA, USA
e-mail: pacton@its.jnj.com



C. Kuntner-Hannes, Y. Haemisch (eds.), *Image Fusion in Preclinical Applications*, https://doi.org/10.1007/978-3-030-02973-9_7

The inflammatory response from an acute stimulus will continue for as long as the stimulus is present. Many of the mediators of inflammation degrade rapidly in tissue. The termination of the acute inflammatory response is vital to prevent tissue damage from the host's own immune system and an accelerating cycle of stimulus and immune response. Resolution of the inflammation is strictly regulated and involves the cessation of the response by the release of anti-inflammatory agents, such as interleukin-10 (IL-10), and the downregulation of pro-inflammatory gene expression. If the stimulus continues for a longer period, a shift to chronic inflammation can be triggered, driving a change in the type of cells present in the target tissue [3]. During chronic inflammation, the healing process occurs, while the damaging effects of the external stimulus are ongoing.

While innate immunity offers security against a wide range of pathogens and other harmful stimuli, a more versatile system, known as adaptive immunity, has evolved to provide increased protection [1]. The adaptive immune system offers an immunological memory after an initial pathogenic invasion, which allows an enhanced response to any subsequent challenges from the same pathogen. In the presence of antigens, such as cell surface proteins on a bacterium, lymphocytes (B- and T-cells) are activated. B-cells secrete antibodies, which are highly selective to the presenting antigen, and prevent binding of the pathogen to host cells. Cytotoxic T-cells induce the death of an invading or tumor cell, while other cell types regulate the complex interaction between the immune system and the pathogen. A class of memory cells provide antigen-specific immunity that persists long after the stimulus has been removed, allowing any subsequent challenge from the same pathogen to be dealt with rapidly. Similar to the innate immune system, careful control of the adaptive immune system is required to ensure the destruction of invading pathogens is complete, while preventing damage to healthy tissue.

Failure of the immune system to manage the delicate cycle between an adequate response to harmful stimuli and removal of that response once the threat has been dealt with can cause a range of inflammatory disorders. Dysregulation of the immune system can lead to an abnormal, aberrant, or out-of-control inflammatory response to external stimuli or even to substances that occur naturally in the body. Allergies are a result of hypersensitivity to an external stimulus, or allergen, leading to an excessive inflammatory response that can be life-threatening [4]. In contrast, autoimmune diseases, such as rheumatoid arthritis (RA) and inflammatory bowel disease (IBD), are caused by the body's immune system attacking its own organs [5]. Indeed, conditions such as obesity, diabetes, and other metabolic diseases, which previously were thought to be unrelated to the immune system, are believed now to be caused by complex interactions between the immune system, adipose tissue, and pancreatic beta cells and can be considered as chronic inflammatory disorders [6–8].

Neuroinflammation represents a special category of immune response. The brain is protected by the blood-brain barrier (BBB) and was considered an immune-privileged organ. However, it is known now that the brain and central nervous system (CNS) are in direct connection with the immune system [9]. Specialized cells, such as microglia, are the brain's own immune cells and act as the main defense against pathological invasion [10]. They are active scavengers of cellular debris, damaged tissue, and neuritic plaques, and, like macrophages, microglia are phagocytic and release cytokines to trigger an inflammatory response. However, similar to the peripheral immune system, chronic neuroinflammation can lead to an uncontrolled destructive microglial response, worsening the underlying tissue damage. A number of neurological and psychiatric disorders are thought now to have a neuroinflammatory pathogenesis, including Parkinson's disease [11], multiple sclerosis [12], Alzheimer's disease [13], and chronic depression [14, 15], and there may be links between peripheral metabolic disorders and central neuroinflammation [16].

## 7.2 Imaging Inflammation

In order to image inflammation, it is necessary to understand the features of the inflammatory response that are amenable to imaging. A summary of these features is presented in Table 7.1.

One of the most fundamental changes that occurs at the site of inflammation during an immune response is a dramatic change in tissue metabolism [17]. The rapid influx of inflammatory cells, coupled with the increase in vascular permeability and blood flow, leads to a significant increase in the demand for nutrients and oxygen. Myeloid cells, such as neutrophils and macrophages, expend large amounts of energy during recruitment to the site of inflammation and during the process of phagocytosis. This energy demand is met through local depletion of nutrients, such as glucose, and increased oxygen consumption. Neutrophils in particular derive much of their energy from glycolysis [18], which makes them suitable targets for metabolic imaging tracers, such as 2-deoxy-2-($^{18}$F)fluoro-D-glucose ($^{18}$F-FDG) [19]. However, $^{18}$F-FDG is non-specific, and any tissue or organ that is highly metabolic, such as the brain or a tumor, will show increased uptake.

The standard of care for imaging infection and inflammation uses radiolabeled autologous leukocytes. This process involves withdrawal and separation of blood cells from the subject, radiolabeling the white blood cells, followed by reinjection of the tagged cells [20]. The radiolabeled cells, typically tagged with $^{99m}$Tc or $^{111}$In, accumulate at the site of inflammation and can be imaged using single-photon emission computed tomography (SPECT) [21, 22]. While it has been suggested that $^{18}$F-FDG positron emission tomography (PET) could supersede SPECT imaging of radiolabeled leukocytes, making use of the superior performance of PET, in many situations questions remain over the accuracy and suitability of $^{18}$F-FDG in this application [23]. However, in vitro radiolabeling of autologous leukocytes with $^{18}$F-FDG, making use of the leukocyte's avidity for glucose, and then injection into the subject and imaging with PET could provide the best of both worlds—the superior performance of PET coupled with the specificity of labeled leukocyte imaging [24].

Increased blood flow and vascular permeability at the site of inflammation, like changes in metabolism, are amenable to imaging using many conventional techniques [25]. Contrast-enhanced (CE) magnetic resonance imaging (MRI) and X-ray computed tomography (CT), with blood pool contrast agents, can demonstrate

**Table 7.1** List of potential targets for imaging inflammation and some probes and imaging modalities that have been used

| Target | Probe(s) and imaging modality | Comments |
|---|---|---|
| Cell metabolism | $^{18}$F-FDG PET | Non-specific—tumors and other highly metabolic tissues also show increased $^{18}$F-FDG uptake |
| Leukocytes | In vitro cell labeling with $^{99m}$Tc or $^{111}$In and SPECT | Low spatial resolution—non-specific uptake in infection as well as inflammation |
| | Ultrasound microbubbles | Low sensitivity |
| Blood flow | CE-CT or CE-MRI | Leakage of contrast into tissue highly visible. DCE-MRI or CT provides quantitative measures of blood flow |
| | DCE-CT or DCE-MRI | |
| | Ultrasound | |
| Edema | Conventional MRI or CT | Hyperintense T2 signal, reduced CT density |
| | DWI | Changes in ADC good indicator of edema |
| Hypoxia | $^{18}$F-FMISO, $^{18}$F-FAZA, $^{18}$F-HX4, $^{64}$Cu-ATSM PET | Slow clearance and non-specific binding of $^{18}$F-FMISO. High radiation dose for $^{64}$Cu |
| | Optical probes sensitive to nitroreductase | |
| Phagocytosis | SPIO particles and MRI | Low sensitivity |
| | $^{19}$F perfluorocarbons and MRI | Virtually no background signal, but very poor sensitivity at clinical field strengths |
| CD8+ T-cells | $^{89}$Zr-Df-IAB22M2C PET and other labeled antibodies, minibodies, etc. | Can detect a few million CD8+ T-cells in preclinical studies. Residualization of $^{89}$Zr can lead to high background signals in clearance organs |
| Various cell surface markers on immune cells (CD40, MHC, CD11b, etc.) | Various labeled antibodies, minibodies, fragments, etc. | Can be used with both optical and PET/SPECT imaging depending on label |
| TSPO | $^{11}$C-(*R*)-PK11195 PET | Low brain uptake, high non-specific binding |
| | $^{18}$F-PBR06, $^{11}$C-PBR28, etc., PET | Higher specific binding, but susceptible to TSPO polymorphism in humans |
| Proteases | ProSense optical | Activatable fluorophore, so light only emitted in presence of target enzyme. Slow kinetics |
| Matrix metalloproteinase (MMP) | MMPSense optical | Activatable fluorophore, so light only emitted in presence of target enzyme. Slow kinetics |
| | PET tracers for MMP | |
| Bone density, volume, surface area, roughness | CT | Structural changes visible on CT often lag behind the inflammatory response |

changes in blood volume around a site of inflammation and any leakage into the interstitial space. Dynamic contrast-enhanced (DCE) imaging allows the generation of quantitative measures of vascular blood flow and leakage. Areas of edema lead to a significant increase in the T2-weighted signal on non-contrast MRI, while CT shows a reduced region of X-ray density. Diffusion-weighted (DW) MRI also exhibits increased apparent diffusion coefficient (ADC), due to the free diffusion of water molecules extravasated from blood vessels in areas of edema.

Inflammation in the brain often leads to disruption of BBB integrity, leading to vascular leakage into the brain. This can be visualized using very similar methods to those utilized for imaging peripheral vascular permeability, including CE-MRI and CE-CT, and DWI. Radioactive tracers also are used to image BBB leakage. Typically, these tracers are not lipid soluble, so will not cross the intact BBB, but are readily extravasated into the brain where the barrier has been compromised by inflammation [25, 26]. In a similar manner, near-infrared (NIR) fluorescent probes have been developed which normally are retained in the blood vessels but will accumulate in the brain in the presence of vascular leakage [27].

Inflamed tissue often becomes hypoxic [28, 29], driven by elevated levels of hypoxia-inducible factors (HIF) [30, 31]. The rapid influx of invading pathogens and immune cells and the release of oxygen-consuming enzymes into an area of inflammation lead to an imbalance in the supply and demand of oxygen, nutrients, and metabolites [32]. While the high metabolic demand of neutrophils makes them a good target for imaging probes such as $^{18}$F-FDG, the presence of hypoxia also provides a suitable target for imaging, and a number of optical and PET probes have been developed which target features of the hypoxic environment. Most often, these probes rely on being taken up and retained by hypoxic cells, while being cleared rapidly by normoxic tissue, giving a high target-to-background ratio.

One of the most widely studied PET imaging agents for hypoxia is $^{18}$F-FMISO [33–35]. Like most of the probes based on nitroimidazoles, this enters cells by passive diffusion, where it is reduced by nitroreductase enzymes. In the presence of normoxic levels of oxygen, the radical is reoxidized rapidly, and free $^{18}$F-FMISO diffuses out of the cell and may be cleared. However, in hypoxic cells, where the partial pressure of oxygen ($pO_2$) is less than approximately 10 mmHg, the reduction products bind to intracellular macromolecules, and the radiolabeled material becomes trapped [36]. While $^{18}$F-FMISO was one of the first PET hypoxia imaging agents, it has drawbacks related to slow clearance and non-specific uptake [37]. Second-generation nitroimidazoles have been developed to improve on the performance of $^{18}$F-FMISO, including $^{18}$F-FAZA and $^{18}$F-HX4 [38]. A number of fluorescent probes make use of the same nitroreductase enzymes to cleave a fluorescent substrate or to trigger Förster resonance energy transfer (FRET), allowing NIR imaging of hypoxia [39–41]. The metal chelate $^{64}$Cu-ATSM [42] also relies on intracellular reduction of the radiolabeled complex, which is reoxidized back to the parent molecule under normoxic conditions. In the presence of hypoxic cells, the reduction products dissociate into $^{64}$Cu(I) which becomes trapped irreversibly inside the cell. Despite a relatively high radiation dose, due to the long half-life of $^{64}$Cu, the imaging performance of $^{64}$Cu-ATSM appears to be superior to $^{18}$F-FMISO [43].

While the majority of imaging studies using these probes have concentrated on tumor hypoxia, the presence of hypoxic inflammatory lesions also should be amenable to this type of imaging [44, 45].

The presence of phagocytes at the site of inflammation is amenable to imaging, by making use of the ability of the cell to take up small particles that are identified as alien to the host. Phagocytosis of small and ultra-small superparamagnetic iron oxide (SPIO) particles by monocytes at the site of inflammation leads to an accumulation of the iron particles, which is visible with MRI. These iron-rich macrophages can be visualized by an increase in image contrast on T1-weighted MRI or a reduction in signal on T2 images [46]. In a similar manner, targeted lipid microbubbles, used for perfusion imaging with ultrasound, can interact with leukocytes, leading to an accumulation in inflamed tissue, which is visible as increased contrast on ultrasound [47].

An alternative to SPIO particles are the perfluorocarbons, which allow an $^{19}$F signal to be detected on a conventional $^{1}$H MRI by retuning the radiofrequency coils [48]. The main advantage of $^{19}$F imaging with MRI is the lack of background signal, due to the absence of naturally occurring $^{19}$F in the soft tissues of the body. In addition, the $^{19}$F signal can be acquired in the same scanning session as the $^{1}$H image, which provides a functional $^{19}$F map overlaid on a spatially aligned anatomical image [49]. Following administration of $^{19}$F-perfluorocarbon, the small emulsion droplets are taken up by phagocytosis into monocytes and macrophages at the site of inflammation [50, 51]. This technique has been used to assay macrophage activity in mouse models of IBD [52], to study treatment response in RA [53], and for measuring peripheral nerve inflammation [54]. The main disadvantage of the $^{19}$F technique is the low sensitivity, particularly at clinical magnetic field strengths, requiring millimolar quantities of contrast agent in each image voxel [49]. This is to be compared with the picomolar concentrations that are detectable with PET.

Phagocytic cells also express surface proteins, which are upregulated in the presence of inflammation. The 18 kDa translocator protein (TSPO), formerly known as the peripheral benzodiazepine receptor, has been identified as a suitable target for imaging inflammation in the periphery and CNS [25]. The radioligand $^{11}$C-(*R*)-PK11195 has been used for PET imaging of TSPO expression for many years [55, 56], although it has relatively poor brain penetration and high non-specific uptake compared to more recent probes [57–59]. These newer tracers include the phenoxyarylacetamides, such as $^{11}$C-DAA1106, and compounds such as $^{11}$C-PBR28 and $^{18}$F-PBR06 (for a review and comparison of these, see [60]). Unfortunately, a problem with these new-generation TSPO PET tracers was identified, in which a polymorphism in the TSPO gene causes changes in binding of the probes [61–63]. However, if each subject is genetically tested for this polymorphism, interpretation of the imaging data may be more accurate.

Proteases are found in all organisms and are responsible for a vast array of processes involving the breakdown of proteins. Specifically for the immune response, proteases are involved in the blood clotting process in response to tissue damage and are active in the complement system [64]. Various proteases are upregulated during inflammation and are a suitable target for imaging probes. Some optical imaging agents make use of the protease enzyme to cleave the probe, separating the fluorophore from the quencher, leading to light emission. These activatable probes are particularly powerful, as they emit light only in the presence of the activating enzyme.

Matrix metalloproteinases (MMPs) are a family of protease enzymes that are involved in a number of cellular processes, including degrading the extracellular matrix, cell proliferation and migration, apoptosis, tissue repair, and immune response. Generally, similar to other proteases, MMPs are upregulated during inflammation and have been used as a target for both optical and radionuclide imaging [65–68]. While early MMP imaging probes were pan-MMP selective, more recently they have been developed to target specific members of the MMP family [69, 70].

The response of the adaptive immune system to pathological invasion results in the influx of large numbers of B- and T-cells to the damaged tissue. These cells typically rely on oxidative phosphorylation as their source of energy, utilizing amino acid, glucose, and lipid metabolism [17]. Many immune cells express cell surface receptors that are amenable to imaging, particularly when the inflammatory process induces a significant upregulation of receptor expression. Cytotoxic T-cells express the CD8 glycoprotein on the cell surface, which is a promising target for imaging T-cell influx and activation at the site of inflammation [71, 72]. Similarly, imaging the expression of CD40, which is upregulated on the surface of many immune cells during inflammation, can be achieved using optical imaging methods [73], while antibody fragments against class II major histocompatibility complex (MHC) and CD11b have been visualized using PET [74].

## 7.3 Imaging Neuroinflammation

Neuroinflammation has been implicated in the pathophysiology of a number of psychiatric and neurological disorders, such as Alzheimer's disease, depression, stroke, and multiple sclerosis. Imaging of the inflammatory process in the brain could provide an early indication of disease or reveal the extent of damage, allowing intervention or treatment [75, 76]. Clinical studies of TSPO expression using PET could provide some of the most valuable insights into neuroinflammation-related diseases. However, many have been inconclusive, where some studies have observed significant increases in TSPO in schizophrenia [77] and depression [78], while others have failed to demonstrate any signs of neuroinflammation [79–82]. The absence of neuroinflammation could be due to the imaging studies being performed too late in the course of the disease, where the inflammatory component already has caused significant loss of brain function, but the underlying inflammation has resolved. Alternatively, neuroinflammation may be present only during extreme periods of symptomatic behavior, such as acute episodes of major depression or psychosis, and imaging TSPO expression during those periods could reveal increased uptake. Similar conflicting results have been observed in neurological disorders, such as Alzheimer's disease [83–86].

Animal models of neuroinflammation have become vital in both the development of imaging probes to study features of the inflammatory process and also in studying the underlying disease and interactions between the peripheral and central immune system [87–89]. Multimodality imaging of neuroinflammation can provide information in individual animals on disruptions to the BBB, molecular indications of the underlying inflammation, such as TSPO, and the expression of inflammatory genes. As an example, Fig. 7.1 shows a preclinical PET-MR study, in which the excitotoxin, kainic acid, was administered peripherally to rats to induce neuroinflammation, and imaging was performed using the TSPO PET tracer $^{18}$F-PBR06 [90–95], and MRI performed to assess cerebral edema. High uptake of $^{18}$F-PBR06 is observed in the thalamus, hippocampus, and amygdala of kainic acid animals, which is absent in healthy controls, and corresponds to regions confirmed in other studies [96]. The brain regions exhibiting increased uptake with PET correspond closely with areas of T2 hyperintensity on MRI, indicating edema (Fig. 7.2). Interestingly, the edema also appears to manifest itself as highly enlarged ventricles, shown clearly as the dark regions of hypointensity on T1 MRI. Many of these features of PET-MR imaging correlate significantly with other indicators of disease

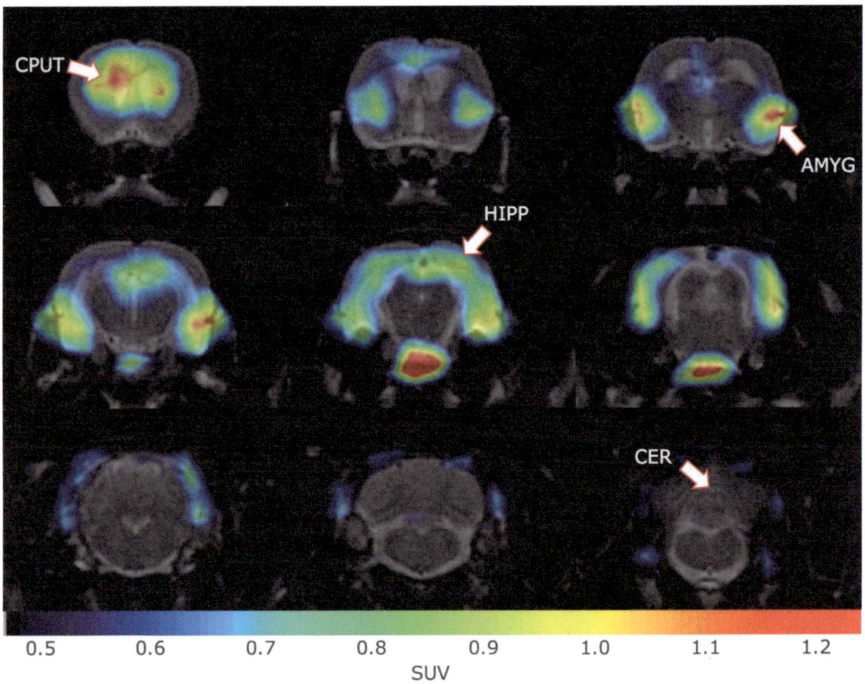


**Fig. 7.1** PET-MR imaging of the TSPO tracer $^{18}$F-PBR06 in a rat model of neuroinflammation, showing increased uptake of the tracer in the brain after kainic acid administration, overlaid on the corresponding MRI slice (*CPUT* caudate-putamen, *AMYG* amygdala, *HIPP* hippocampus, *CER* cerebellum)

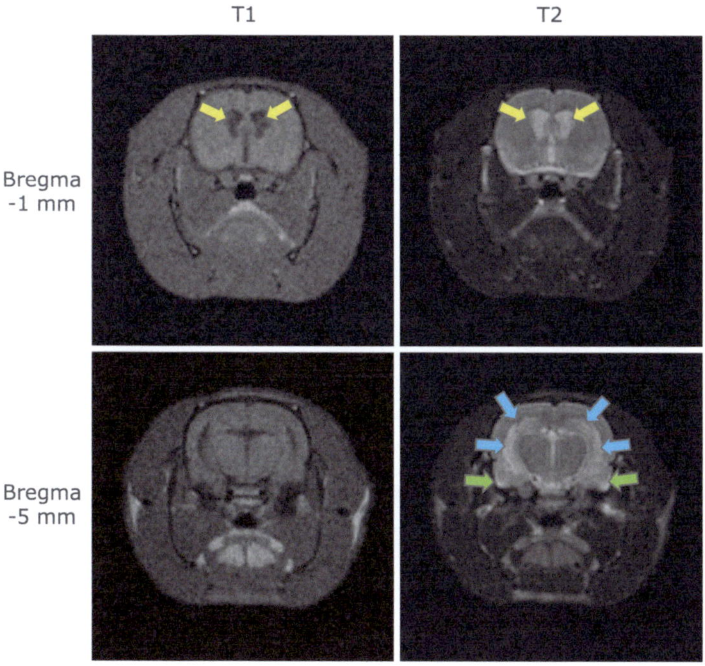


**Fig. 7.2** MRI in a kainic acid rat model of neuroinflammation demonstrates hyperintensity on T2-weighted scans, corresponding to areas of cerebral edema and enlarged ventricles. Yellow arrows indicate fluid accumulation in the ventricles, leading to enlargement. Blue and green arrows show T2 hyperintensity in the hippocampus and amygdala, respectively

severity, such as weight loss. High-field MRI volumetric studies have demonstrated similar results, where kainic acid-treated animals had a significantly smaller hippocampus and increased ventricle size [97].

One significant advantage small animal studies have over clinical applications in humans is the ability to use transgenic animals, in which one or more genes are altered to provide detailed information on inflammatory pathways. A particularly useful mouse model developed recently uses a firefly luciferase (luc) reporter gene, coupled to a 12 kb fragment of the glial fibrillary acidic protein (GFAP) promoter and a 850 bp human β-globin intron 2, to allow bioluminescence imaging of GFAP expression [98]. The GFAP protein is expressed by a number of cell types in the CNS, including astrocytes, and is known to be significantly increased during neuroinflammation. Indeed, GFAP immunohistochemistry is one of the standard methods for studying astrocyte activation in the presence of neurotoxins, neuronal injury, and inflammation, although the precise mechanism of GFAP upregulation is unclear.

The GFAP-luc mouse (FVB/N-Tg(Gfap-luc)53Xen) has been studied extensively in a number of inflammatory models [90, 98–101]. Peripheral administration of pro-inflammatory agents, such as lipopolysaccharide (LPS), kainic acid, and

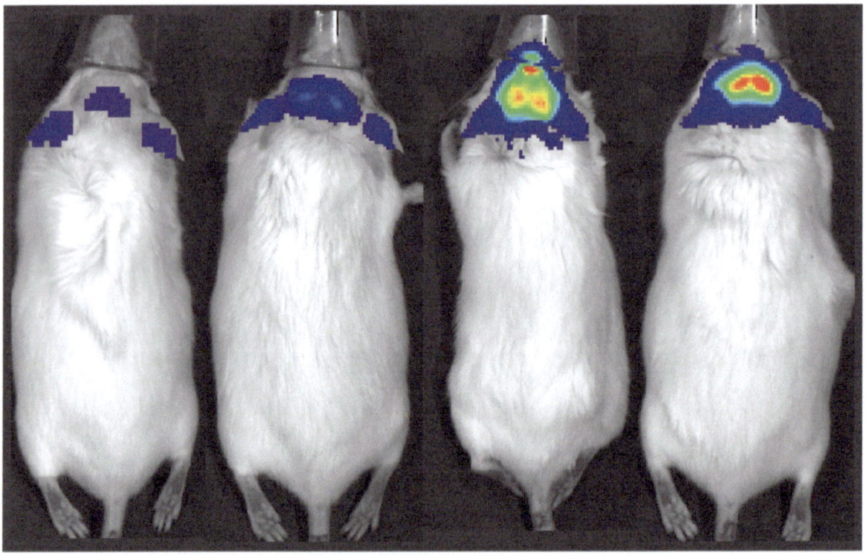


**Fig. 7.3** Imaging bioluminescence light emission from Gfap-luc transgenic mice after administration of increasing doses of the pro-inflammatory compound kainic acid

TNFα, leads to increased expression of the GFAP gene and the emission of bioluminescence light in the presence of the luciferin substrate (Fig. 7.3). The amount of light emitted increases by up to two orders of magnitude from baseline and exhibits a steep dose-response (Fig. 7.4), which is highly correlated with seizure score and other measures of sickness behavior, such as weight loss. The time course of light emission after a single dose of kainic acid exhibits a rapid increase up to 24 h post-administration, which remains stable for several days and then begins to resolve slowly after 3–4 days (Fig. 7.5), which is mirrored in measures such as weight loss.

One consequence of such a steep dose-response could be the impact on therapeutic studies. At the higher doses of kainic acid, the induced inflammation and neuronal damage could be so severe that it would be impossible for any therapeutic to recover the lost brain function. Therefore, using lower doses, around the region of rapid increase of the dose-response curve, could be more useful in studies of treatment effect. With a dose of 20 mg/kg kainic acid, a study was able to demonstrate inhibition of astrogliosis in the GFAP-luc mouse model with treatment by colony-stimulating factor 1 (CSF1), but only when treated prior to, or up to 6 h after, kainic acid [102]. Any further delay in treatment, beyond 6 h, could no longer rescue damaged neurons.

It should be noted also that the sensitivity to pro-inflammatory agents, such as kainic acid, is highly age- and strain-dependent for both rats [103] and mice [104, 105]—for example, DBA/2J mice are very sensitive to chemically induced seizures, while C57BL/6J are quite resistant.

**Fig. 7.4** Dose-response of bioluminescence light emission from the brain of Gfap-luc transgenic mice after administration of kainic acid. Dose range of kainic acid is 5–30 mg/kg, and $ED_{50}$ (the dose at which 50% of the maximum response is observed) is 10.6 ± 0.2 mg/kg

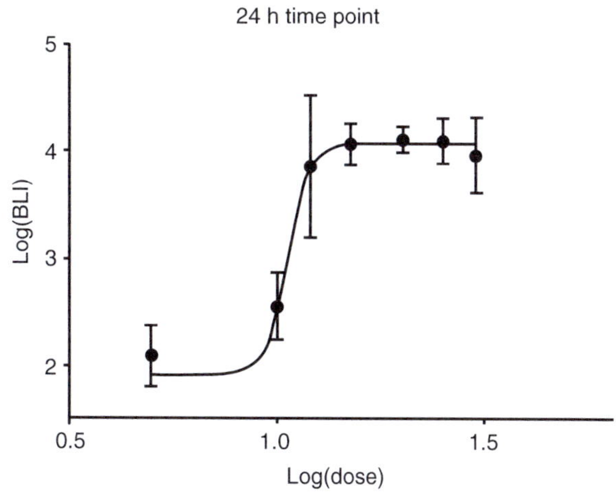


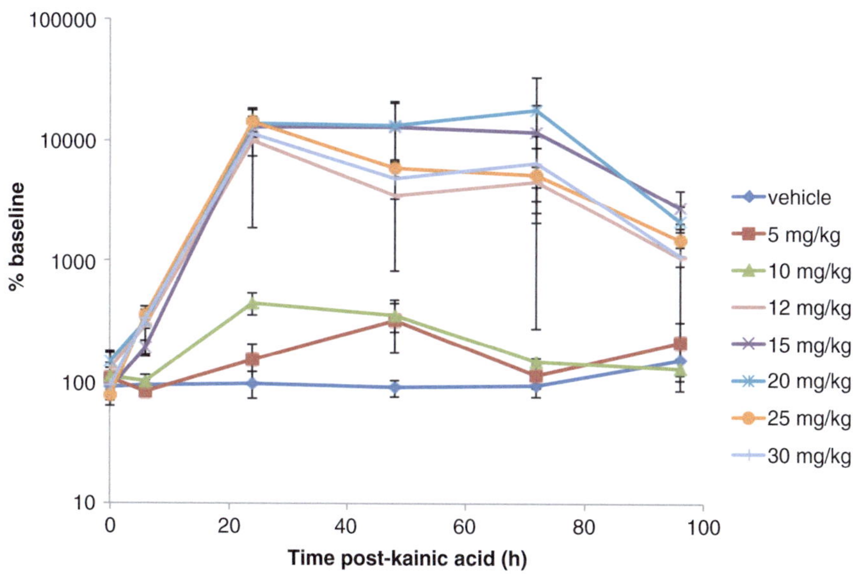


**Fig. 7.5** Time course of bioluminescence light emission after acute administration of varying doses of kainic acid

## 7.4 Imaging Rheumatoid Arthritis

Rheumatoid arthritis (RA) is a debilitating chronic autoimmune disease that affects the joints, resulting in pain, stiffness, and swelling [106]. Inflammation of the synovial membrane eventually leads to tendon tethering and destruction of the cartilage and bone surface in the joint, causing loss of function and deformity.

While primarily a disease of the joint, RA can lead to reactions in the skin, lungs, kidneys, heart, eyes, and other organs. Diagnosis of RA typically is done using conventional imaging methods, such as X-rays of the affected joints looking for swelling of the soft tissue and early bone erosion. However, structural changes due to RA, which are visible on anatomical imaging modalities, often occur long after the underlying chronic inflammation has taken hold. The use of molecular imaging methodologies, looking for the specific signatures of inflammation, may enable much earlier treatment, before joint damage has occurred [107–109]. Similarly, detailed analysis of DCE-MRI and diffusion tensor imaging (DTI), coupled with the use of novel contrast agents, could provide an early signal from changes in the synovial space [110, 111]. Ultrasound offers similar opportunities for imaging the structure of the affected joint and changes in blood flow resulting from the inflammatory response [112].

A number of genetic and environmental factors may play a role in the pathophysiology of RA, although the precise mechanism of the initial cellular activation is unknown [113]. The initiating inflammatory trigger leads to a T-cell-mediated amplification of the inflammation and a chronic phase leading to joint injury and bone erosion. Cytokines such as IL-1, TNFα, and IL-6 are released at high concentrations into the synovial fluid, which accelerate the tissue damage. TNFα appears to be one of the main pro-inflammatory cytokines in RA, which has led to the development of a number of therapeutics targeting this chemical, primarily anti-TNFα antibodies and fusion proteins. However, since TNFα is known to inhibit tumorigenesis and viral replication, the inhibition of this cytokine can have serious side effects, such as certain cancers and a vulnerability to opportunistic infection and tuberculosis (TB). While biologic therapeutics have been successful in treating RA, these potential side effects from systemic exposure can lead to therapy discontinuation and, in the most severe cases, death. Therefore, strategies to develop more targeted therapies to the joint, while reducing systemic toxicity, have been developed, and imaging will play a vital role in measuring joint penetration of these treatments [114].

Despite their inherent limitations, animal models of RA have been vital to progress our understanding of the disease and in the development of new therapeutics [115, 116]. Collagen-induced arthritis (CIA) is one of the most widely used models, utilizing immunization of animals with type II collagen to induce symptoms very similar to human RA. Histological examination reveals an influx of cells into synovial tissue that resembles RA and destruction of bone and cartilage. While disease penetration is strain-dependent, one advantage of the CIA model is that it can be used not only in rodents but also in nonhuman primates (NHP) [117].

Imaging has been used extensively in the CIA model, using methods to probe the underlying inflammation and also structural imaging of bone loss. Although a nonspecific marker of the inflammatory process, $^{18}$F-FDG PET is a useful method to quantify the increased metabolic activity induced by joint inflammation [118]. Uptake of $^{18}$F-FDG in the joint is correlated with clinical and histopathological scores of disease severity throughout the onset and progression of disease [119].

Dual-probe fluorescence imaging in mice has allowed the simultaneous acquisition of signals from MMP activity, using the commercially available probe MMPSense, and macrophage influx, using a fluorescently tagged antibody to CD11b [120]. The response to treatment can be monitored using molecular imaging of a number of inflammatory targets, including fibroblast activation protein (FAP), macrophages, and $\alpha_v\beta_3$ integrins [121, 122]. Similarly, $^{19}$F-MRI of perfluorocarbons in the joint of CIA rats is correlated with disease severity and is an indicator of response to therapy [53]. Combining imaging modalities allows the greatest flexibility, providing imaging of joint metabolism with $^{18}$F-FDG PET and optical imaging of NIR fluorescent probes to visualize targets such as MMP, proteases, cathepsins, and others [123]. Figure 7.6 demonstrates the application of PET and optical imaging of the hind paw in a CIA mouse model of RA. Interestingly, $^{18}$F-FDG uptake can be measured using conventional PET imaging and also by detecting the Cerenkov light emission produced by the positron [124, 125].

PET imaging of $^{18}$F-fluoride uptake by bone provides valuable information on bone mineral deposition and osteoblastic activity. Although used primarily for detection and quantification of metastatic bone cancer, fluoride PET can detect murine arthritis as early as 14 days after administration of pro-inflammatory agents [126]. Comparison with CT of the same joints revealed a reasonable correlation between clinical scores or bone surface measurements and $^{18}$F-fluoride uptake.

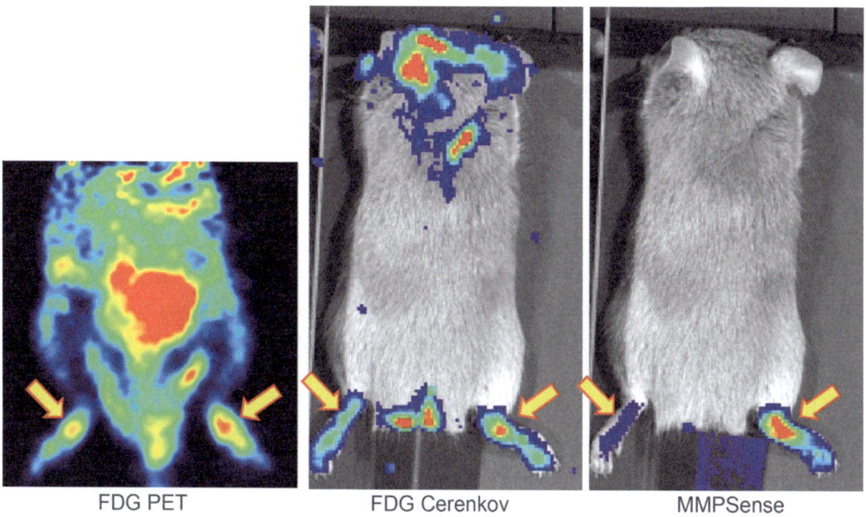


**Fig. 7.6** Imaging inflammation in the joints of CIA mice (arrowed). $^{18}$F-FDG uptake can be observed using both conventional PET imaging and optical imaging of Cerenkov light emission. In the same animals, MMP expression is measured using MMPSense and NIR fluorescence imaging. Note the field of view of the PET scan is smaller than the other optical images, and does not cover the whole mouse. Also, the right paw exhibits increased FDG and MMPSense uptake in all images, indicating a greater degree of inflammation in that paw compared to the other

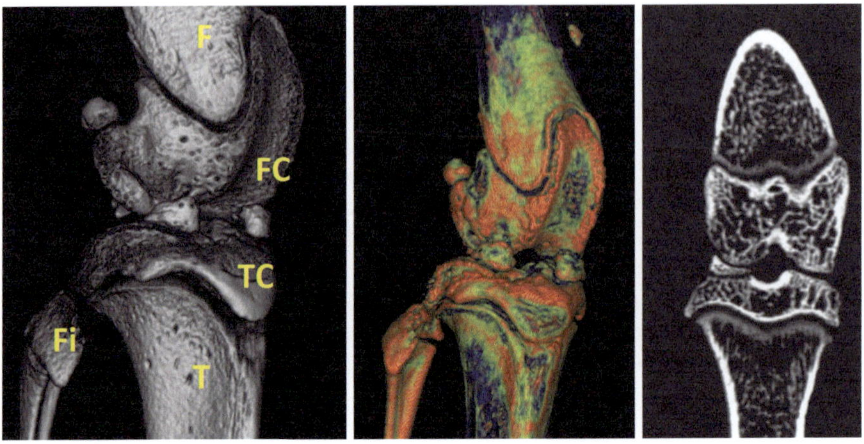


**Fig. 7.7** High-resolution CT imaging of a hind paw demonstrates dramatic destruction of bone in the CIA mouse, leading to decreased bone density and surface erosion (left). A map of the surface roughness indicates the joint regions most affected by the disease (center), while detailed examination of the joint reveals bone thinning at the growth plate boundary and loss of cartilage and collapse of joint space (right). *F* femur, *T* tibia, *Fi* fibula, *FC* condyle of femur, *TC* condyle of tibia

Bone erosion in animal models of RA can be visualized using high-resolution CT imaging of the joint (Fig. 7.7). While a general decrease in bone density in the affected regions is expected, other features, such as joint space reduction due to cartilage damage and changes in bone volume and surface area, also are resolved quite clearly [127, 128]. In addition, quantification of bone erosion has been performed by measuring the roughness of the bone surface in the vicinity of the joint [129, 130]. This roughness is defined in a number of ways, including the depth of features from a plane parallel to the local surface or from the angle of intercepting polygons drawn on a surface rendering. The distribution of roughness values demonstrates features consistent with the progression of disease, such as a shift to higher absolute roughness values in more severe disease (Fig. 7.8).

The potentially severe side effects from conventional treatments for RA, such as anti-TNFα antibodies, have driven the search for more targeted therapies, which will maximize the concentration of the drug at the affected joint, while minimizing systemic exposure and the resulting toxicity [114]. Most of the targeted therapies make use of features of the arthritic joint, such as increased vascular permeability, changes in lymphatic drainage, alterations in temperature or pH, and increased expression of certain enzymes or other proteins. A variety of novel drug technologies have been employed, including bispecific biologics, targeted nanoparticles, and polymer-drug conjugates. Radiolabeling these therapeutics, and imaging with PET, allows the quantitative measurement of uptake in a rodent model of RA, demonstrating improved joint targeting (Fig. 7.9).

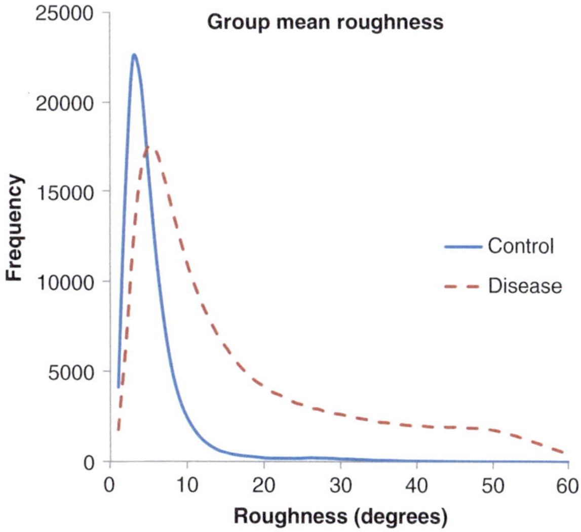


**Fig. 7.8** Analysis of the surface roughness distribution in CIA mice shows the shift of the distribution to the right, indicating more severe disease and bone destruction

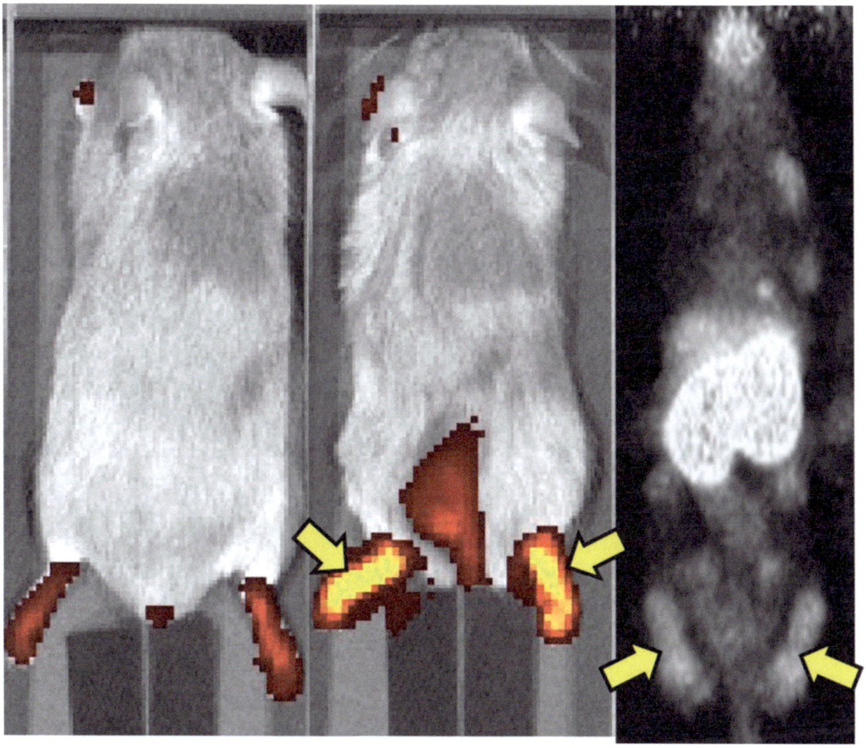

**Fig. 7.9** PET imaging of a radiolabeled therapeutic demonstrates enhanced uptake into the affected joint in RA (right picture, arrowed), which is confirmed on a fluorescently tagged version of the same compound (center picture, arrowed). A healthy control animal exhibits significantly lower uptake into the joint (left picture)

## 7.5 Imaging Inflammatory Bowel Disease

Inflammatory bowel disease (IBD) comprises both Crohn's disease and ulcerative colitis. IBD is characterized by inflammation of the ileum, colon, and rectum and can involve both mucosal and transmural inflammation. It is believed that a genetic predisposition to IBD leads to an inappropriate or unregulated response to the intestinal microbiome, although there also is a strong environmental component of the disease relating to diet, smoking behavior, and use of certain medicines, particularly antibiotics [131, 132]. IBD is a known risk factor for the development of colorectal cancer [133, 134], and, as such, methods for imaging inflammation in the colon could provide an early indication of potential tumor development [135].

Many of the conventional techniques for imaging inflammation, such as MRI, CT, CE-MRI, CE-CT, DWI, $^{18}$F-FDG PET-CT, and autologous leukocytes, are applicable in IBD [136–146]. MRI and CT contrast agents administered enterically increase the contrast between the lumen and bowel wall, allowing better visualization of bowel wall thickening, abscesses, edema, and mucosal enhancement. Like in other inflammatory diseases, $^{18}$F-FDG PET can be used to study IBD, making use of the influx of inflammatory cells and increased cellular metabolism at the site of local disease. Multimodality PET-CT imaging is vital in this application to identify accurately the location, by CT, of any inflammatory lesions seen on PET. However, the non-specific nature of $^{18}$F-FDG could lead to the misidentification of a potential tumor as an IBD lesion, making this technique less useful for primary diagnosis.

IBD and colon cancer offer an almost unique opportunity for imaging, since the location of the inflammation is amenable to endoscopic examination, and could be included as part of a routine colonoscopy [147]. This opens up the possibility of using a variety of optical imaging techniques, which would be impossible in most other organs due to the depth of tissue being imaged, and for topical administration of any optical contrast agents [148–151]. Some methodologies used in endoscopic imaging of the GI tract include conventional white light endoscopy, NIR fluorescence imaging, confocal laser endomicroscopy, Raman endoscopy, optical coherence tomography, narrow-band imaging, and hyperspectral imaging [152–155]. Advances in nanotechnology have led to the development of targeted nanoparticles that carry a diagnostic or therapeutic payload. Tagging these nanoparticles with a fluorescent dye or radioisotope allows in vivo imaging of cellular and molecular processes occurring in IBD [156].

A number of animal models of IBD have been developed that provide valuable information on the pathophysiology of intestinal inflammation and the mechanisms that drive loss of mucosal and epithelial barrier homeostasis [157–160]. One of the most common chemically induced models involves the administration of the chelating agent dextran sulfate sodium (DSS) in the drinking water of mice. This causes acute inflammation of the colon, leading to weight loss, diarrhea, and rectal bleeding. A number of genetically engineered models of IBD also have been developed, including those with innate immunity defects or aberrant adaptive immune cell response [161, 162]. However, there are vast differences between these mouse models and human IBD, particularly as the human disease is driven by a wide array of genetic and environmental factors, while experimental models tend to result from a single genetic deletion or transgene overexpression. Therefore, these mouse models should be used with caution in the development of novel therapeutics, as they cannot represent the complex genetic, immune system, and environmental interactions seen in humans [163].

Imaging studies of preclinical models of IBD have utilized a variety of imaging modalities and probes to understand the inflammatory response. A study to compare a number of different NIR fluorescent probes in a DSS mouse model of colitis found that most of the activatable probes that are sensitive to protease and MMP enzymes gave a good signal in areas of inflammation [164]. While the fluorescence signal was correlated with histology and colitis scores, the high background signals may still provide a problem during in vivo optical imaging. Similarly, a study using a targeted γ-glutamyltranspeptidase (GGT) probe, which is activated in the presence of the GGT enzyme, demonstrated high uptake in a DSS mouse model [165]. The advantage of this optical imaging agent was that the probe was delivered topically, via a spray-on method, significantly reducing systemic exposure and any potential toxicity.

An extensive comparison of different imaging modalities and probes was performed in the DSS colitis model, which revealed that many of the conventional clinical imaging methods are equally applicable to mouse models [137]. PET imaging of DSS mice has been performed to compare the non-specific metabolic tracer, $^{18}$F-FDG (Fig. 7.10), with the TSPO imaging agent, $^{18}$F-DPA-714 [166]. While $^{18}$F-FDG has been shown to be useful for quantifying colon cancer [167], the inflammation from acute DSS administration was found to be lower, and not accurately resolved with $^{18}$F-FDG. However, TSPO imaging demonstrated significantly increased uptake of the tracer, which could provide a more sensitive measure to monitor the progression of disease and response to anti-inflammatory treatment. Other PET imaging agents specific to targets known to be increased in IBD also could have value, such as the inducible form of the cyclooxygenase enzyme, COX-2 [168].

Similar to RA, anti-TNFα therapies have been effective in treating some patients with IBD. However, not all patients respond to anti-TNFα treatment, which could be due to variations in the expression of the target in the bowel. Molecular imaging of TNFα expression prior to treatment, using a fluorescently tagged antibody to membrane-bound TNFα, has demonstrated that patients with high levels of TNFα show significantly improved response to treatment and has the potential to predict response and select suitable patients for this type of treatment [169]. Methods to deliver targeted therapies to the GI tract for IBD would have significant value in reducing the systemic exposure and associated toxicities [170–173]. Visualization of the delivery of the drug with imaging would be vital to ensure it reaches all affected parts of the ileum, colon, and rectum [174]. Indeed, contrast agents

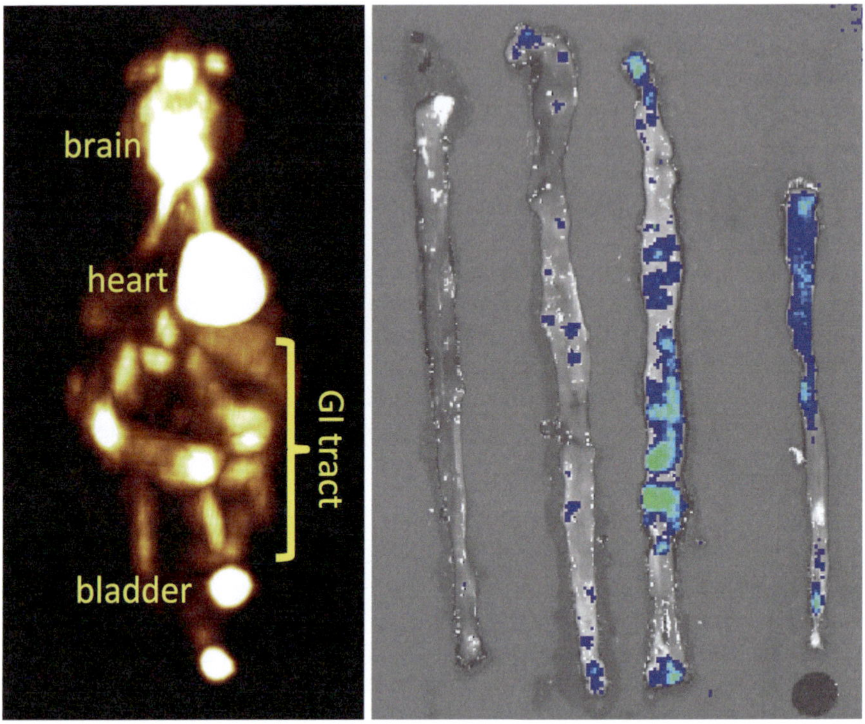


**Fig. 7.10** $^{18}$F-FDG uptake in a DSS mouse model of IBD, showing increased uptake in the GI tract (left), which is confirmed by ex vivo Cerenkov imaging (right). The four GI tract sections shown are (left to right) the duodenum, jejunum, ileum, and colon (cecum has been removed). Image courtesy of Dr. Peter King, Janssen Research and Development

encapsulated into the same targeted delivery technology as the therapeutic combines both treatment and imaging in a single package [175].

## 7.6 Conclusions

Inflammation is implicated in the pathophysiology of a vast range of diseases, from Alzheimer's disease and depression to chronic autoimmune disorders such as RA and IBD, and even diabetes and other metabolic disorders. Imaging the inflammatory response and immune system is becoming increasingly important in the understanding of these disorders and for the development of novel therapeutics. Many targets associated with inflammation are amenable to imaging, including nonspecific changes in metabolic activity or blood flow, alterations in enzymatic activity, and expression of cell surface markers. However, multimodality imaging, incorporating features of both anatomical and molecular imaging, targeting a variety of processes involved in the inflammatory response will maximize our understanding of these disorders.

## References


1. Murphy K, Weaver C. Janeway's Immunobiology. 9th ed. New York: Garland Science; 2016.
2. Kolaczkowska E, Kubes P. Neutrophil recruitment and function in health and inflammation. Natl Rev. 2013;13:159–75.
3. Murakami M, Hirano T. The molecular mechanisms of chronic inflammation development. Front Immunol. 2012;3:323.
4. O'Hehir RE, Holgate ST, Sheikh A. Middleton's allergy essentials. San Diego: Elsevier; 2016.
5. Mackay IR, Rose NR, editors. The autoimmune diseases. San Diego: Elsevier; 2014.
6. Donath MY, Shoelson SE. Type 2 diabetes as an inflammatory disease. Nat Rev Immunol. 2011;11:98–107.
7. Gonzalez-Chavez A, et al. Pathophysiological implications between chronic inflammation and the development of diabetes and obesity. Cir Cir. 2011;79:190–7.
8. Lumeng CN, Saltiel AR. Inflammatory links between obesity and metabolic disease. J Clin Invest. 2011;121:2111–7.
9. Ader R, Felten D, Cohen N. Interactions between the brain and the immune system. Annu Rev Pharmacol Toxicol. 1990;30:561–602.
10. Nayak D, Roth TL, McGavern DB. Microglia development and function. Annu Rev Immunol. 2014;32:367–402.
11. Vivekanantham S, Shah S, Dewj R, et al. Neuroinflammation in Parkinson's disease: role in neurodegeneration and tissue repair. Int J Neurosci. 2015;125:717–25.
12. Ellward E, Zipp F. Molecular mechanisms linking neuroinflammation and neurodegeneration in MS. Exp Neurol. 2014;262:8–17.
13. Heneka MT, Carson MJ, El Khoury J, et al. Neuroinflammation in Alzheimer's disease. Lancet Neurol. 2015;14:388–405.
14. Bhattacharya A, Derecki NC, Lovenberg T, et al. Role of neuro-immunological factors in the pathophysiology of mood disorders. Psychopharmacologia. 2016;233:1623–36.
15. Dantzer R, O'Connor JC, Freund GG, et al. From inflammation to sickness and depression: when the immune system subjugates the brain. Nat Rev Neurosci. 2008;9:46–56.
16. Van Dijk G, van Heijnengen S, Reijne AC, et al. Integrative neurobiology of metabolic diseases, neuroinflammation, and neurodegeneration. Front Neurosci. 2015;9:173. https://doi.org/10.3389/fnins.2015.00173.
17. Kominsky DJ, Campbell EL, Colgan SP, et al. Metabolic shifts in immunity and inflammation. J Immunol. 2010;184:4062–8.
18. Borregaard N, Herlin T. Energy metabolism of human neutrophils during phagocytosis. J Clin Invest. 1982;70:453–65.
19. Wu C, Li F, Niu G, et al. PET imaging of inflammation biomarkers. Theranostics. 2013;3:448–66.
20. Sampson CB. Labeled cells for imaging infection. In: Cox PH, Buscombe JR, editors. The imaging of infection and inflammation. Heidelberg: Springer; 1998. p. 31–60.
21. Rodie ME. Imaging inflammation with Tc-99m hexamethyl propylene amine oxime (HMPAO) labeled leucocytes. Radiology. 1988;166:767–72.
22. Thakur ML. Indium-111 labeled leukocytes for the localization of abscesses: preparation, analysis, tissue distribution and comparison with gallium-67 citrate in dogs. Lab Clin Med. 1977;89:217–28.
23. Glaudemans AWJM, Signore A. FDG-PET/CT in infections: the imaging method of choice? Eur J Nucl Med Mol Imaging. 2010;37:1986–91.
24. Forstrom LA, Mullan BP, Hung JC, et al. $^{18}$F-FDG labeling of human leukocytes. Nucl Med Commun. 2000;21:691–4.
25. Jacobs AH, Tavitian B, INMiND Consortium. Noninvasive molecular imaging of neuroinflammation. J Cereb Blood Flow Metab. 2012;32:1393–415.
26. Wunder A, Klohs J, Dirnagl U. Non-invasive visualization of CNS inflammation with nuclear and optical imaging. Neuroscience. 2009;158:1161–73.

27. Klohs J, Steinbrink J, Bourayou R, et al. Near-infrared fluorescence imaging with fluorescently labeled albumin: a novel method for non-invasive optical imaging of blood-brain barrier impairment after focal cerebral ischemia in mice. J Neurosci Methods. 2009;180:126–32.
28. Biddlestone J, Bandarra D, Rocha S. The role of hypoxia in inflammatory disease (Review). Int J Mol Med. 2015;35:859–69.
29. Eltzschig HK, Carmeliet P. Hypoxia and inflammation. N Engl J Med. 2011;364:656–65.
30. Colgan SP, Campbell EL, Kominsky DJ. Hypoxia and mucosal inflammation. Annu Rev Pathol. 2016;11:77–100.
31. Cummins EP, Keogh CE, Crean D, et al. The role of HIF in immunity and inflammation. Mol Asp Med. 2016;47-48:24–34.
32. Koeppen M, Eckle T, Eltzschig HK. The hypoxia-inflammation link and potential drug targets. Curr Opin Anaesthesiol. 2011;24:363–9.
33. Dubois L, Landuyt W, Haustermans K, et al. Evaluation of hypoxia in an experimental rat tumor model by [(18)F]fluoromisonidazole PET and immunohistochemistry. Br J Cancer. 2004;91:1947–54.
34. Prekeges JL, Rasey JS, Grunbaum Z, et al. Reduction of fluoromisonidazole, a new imaging agent for hypoxia. Biochem Pharmacol. 1991;42:2387–95.
35. Tang G, Wang M, Tang X, et al. Fully automated one-pot synthesis of [(18)F]fluoromisonidazole. Nucl Med Biol. 2005;32:553–8.
36. Yip C, Blower PJ, Goh V, et al. Molecular imaging of hypoxia in non-small cell lung cancer. Eur J Nucl Med Mol Imaging. 2015;42:956–76.
37. Krohn KA, Link JM, Mason RP. Molecular imaging of hypoxia. J Nucl Med. 2008;49(suppl 2):129S–48S.
38. Peeters SGJA, Zegers CML, Lieuwes NG, et al. A comparative study of the hypoxia PET tracers [$^{18}$F]HX4, [$^{18}$F]FAZA, and [$^{18}$F]FMISO in a preclinical tumor model. Int J Radiat Oncol Biol Phys. 2015;91:351–9.
39. Guo T, Cui L, Shen J, et al. A highly sensitive long-wavelength fluorescence probe for nitroreductase and hypoxia: selective detection and quantification. Chem Commun. 2013;49:10820–2.
40. Kiyose K, Hanaoka K, Oushiki D, et al. Hypoxia-sensitive fluorescent probes for *in vivo* real-time fluorescence imaging of acute ischemia. J Am Chem Soc. 2010;132:15846–8.
41. Xu K, Wang F, Pan X, et al. High selectivity imaging of nitroreductase using near-infrared fluorescence probe in hypoxic tumor. Chem Commun. 2013;49:2554–6.
42. Vavere AL, Lewis JS. Cu-ATSM: a radiopharmaceutical for PET imaging of hypoxia. Dalton Trans. 2007;43:4893–902.
43. Bourgeois M, Rajerison H, Guerard F, et al. Contribution of [$^{64}$Cu]-ATSM PET in molecular imaging of tumor hypoxia compared to classical [$^{18}$F]-MISO—a selected review. Nucl Med Rev Cent East Eur. 2011;14:90–5.
44. Buscombe JR. Exploring the nature of atheroma and cardiovascular inflammation *in vivo* using positron emission tomography (PET). Brit J Radiol. 2015;88:20140648.
45. Liu R-S, Chou T-K, Chang C-H, et al. Biodistribution, pharmacokinetics and PET imaging of [$^{18}$F]FMISO, [$^{18}$F]FDG and [$^{18}$F]FAc in a sarcoma- and inflammation-bearing mouse model. Nucl Med Biol. 2009;36:305–12.
46. Stoll G, Bendszus M. Imaging of inflammation in the peripheral and central nervous system by magnetic resonance imaging. Neuroscience. 2008;158:1151–60.
47. Lindner JR, Song J, Xu F, et al. Noninvasive ultrasound imaging of inflammation using microbubbles targeted to activated leukocytes. Circulation. 2000;102:2745–50.
48. Ruiz-Cabello J, Barnett BP, Bottomley PA, et al. Fluorine ($^{19}$F) MRS and MRI in biomedicine. NMR Biomed. 2011;24:114–29.
49. Hu L, Hockett FD, Chen J, et al. A generalized strategy for designing $^{19}$F/$^{1}$H dual-frequency MRI coil for small animal imaging at 4.7 Tesla. J Magn Reson Imaging. 2011;34:245–52.
50. Ahrens ET, Young W-B, Xu H, et al. Rapid quantification of inflammation in tissue samples using perfluorocarbon emulsion and fluorine-19 nuclear magnetic resonance. BioTechniques. 2011;50:229–34.

51. Jacoby C, Temme S, Mayenfels F, et al. Probing different perfluorocarbons for in vivo inflammation imaging by 19F MRI: image reconstruction, biological half-lives and sensitivity. NMR Biomed. 2014;27:261–71.
52. Kadayakkara DK, Ranganathan S, Young WB, et al. Assaying macrophage activity in a murine model of inflammatory bowel disease using fluorine-19 MRI. Lab Investig. 2012;92:636–45.
53. Balducci A, Helfer BM, Ahrens ET, et al. Visualizing arthritic inflammation and therapeutic response by fluorine-19 magnetic resonance imaging ($^{19}$F MRI). J Inflamm. 2012;9:24. https://doi.org/10.1186/1476-9255-9-24.
54. Weise G, Basse-Luesebrink TC, Wessig C, et al. In vivo imaging of inflammation in the peripheral nervous system by $^{19}$F MRI. Exp Neurol. 2011;229:494–501.
55. Pike VW, Halldin C, Crouzel C, et al. Radioligands for PET studies of central benzodiazepine receptors and PK (peripheral benzodiazepine) binding sites—current status. Nucl Med Biol. 1993;20:503–25.
56. Venneti S, Lopresti BJ, Wiley CA. The peripheral benzodiazepine receptor (translocator protein 18kDa) in microglia: from pathology to imaging. Prog Neurobiol. 2006;80:308–22.
57. Ching ASK, Kuhnast B, Damont A, et al. Current paradigm of the 18-kDa translocator protein (TSPO) as a molecular target for PET imaging in neuroinflammation and neurodegenerative diseases. Insights Imaging. 2012;3:111–9.
58. Turkheimer FE, Rizzo G, Bloomfield PS, et al. The methodology of TSPO imaging with positron emission tomography. Biochem Soc Trans. 2015;43:586–92.
59. Vivash L, O'Brien TJ. Imaging microglial activation with TSPO PET: lighting up neurological disease? J Nucl Med. 2016;57:165–8.
60. Chauveau F, Boutin H, Van Camp N, et al. Nuclear imaging of neuroinflammation: a comprehensive review of [$^{11}$C]PK11195 challengers. Eur J Nucl Med Mol Imaging. 2008;35:2304–19.
61. Kreisl WC, Fujita M, Fujimara Y, et al. Comparison of [$^{11}$C]-(*R*)-PK 11195 and [$^{11}$C]PBR28, two radioligands for translocator protein (18 kDa) in human and monkey: implications for positron emission tomographic imaging of this inflammation biomarker. NeuroImage. 2009;49:2924–32.
62. Owen DR, Yeo AJ, Gunn RN, et al. An 18-kDa translocator protein (TSPO) polymorphism explains differences in binding affinity of the PET radioligand PBR28. J Cereb Blood Flow Metab. 2012;32:1–5.
63. Rojas C, Stathis M, Coughlin JM, et al. The low-affinity binding of second generation radiotracers targeting TSPO is associated with a unique allosteric binding site. J Neuroimmune Pharmacol. 2018;13:1–5.
64. Bunnett NW. Protease-activated receptors: how proteases signal to cells to cause inflammation and pain. Semin Thromb Hemost. 2006;32(Suppl 1):39–48.
65. Klohs J, Baeva N, Steinbrink J, et al. In vivo near-infrared fluorescence imaging of matrix metalloproteinase activity after cerebral ischemia. J Cereb Blood Flow Metab. 2009;29:1284–92.
66. Leahy AA, Esfahani SA, Foote AT, et al. Following the trajectory of osteoarthritis development through serial near infrared fluorescence imaging of MMP activities. Arthritis Rheum. 2015;67:442–53.
67. Matusiak N, Waarde A, Bischoff R, et al. Probes for non-invasive matrix metalloproteinase-targeted imaging with PET and SPECT. Curr Pharmaceutical Design. 2013;19:4647–72.
68. Wagner S, Breyholz H-J, Faust A, et al. Molecular imaging of matrix metalloproteinases in vivo using small molecule inhibitors for SPECT and PET. Curr Med Chem. 2006;13:2819–38.
69. Bordenave T, Helle M, Beau F, et al. Synthesis and in vitro and in vivo evaluation of MMP-12 selective optical probes. Biconjugate Chem. 2016;27(10):2407–17. https://doi.org/10.1021/acs.bioconjchem.6b00377.
70. Qin H, Zhao Y, Zhang J, et al. Inflammation-targeted gold nanorods for intravascular photoacoustic imaging detection of matrix metalloproteinase-2 (MMP2) in atherosclerotic plaques. Nanomedicine. 2016;12:1765–74.
71. Tavare R, McCracken MN, Zettlitz KA, et al. Engineered antibody fragments for immune-PET imaging of endogenous CD8+ T cells in vivo. Proc Natl Acad Sci. 2014;111:1108–13.

72. Tavare R, Escuin-Ordinas H, Mok S, et al. An effective immune-PET imaging method to monitor CD8-dependent responses to immunotherapy. Cancer Res. 2015;76:73–82.
73. Klohs J, Grafe M, Graf K, et al. In vivo imaging of the inflammatory receptor CD40 after cerebral ischemia using a fluorescent antibody. Stroke. 2008;39:2845–52.
74. Rashidian M, Keliher EJ, Bilate AM, et al. Noninvasive imaging of immune response. Proc Natl Acad Sci. 2015;112:6146–51.
75. Albrecht DS, Granziera C, Hooker JM, et al. In vivo imaging of human neuroinflammation. ACS Chem Neurosci. 2016;7:470–83.
76. Pulli B, Chen JW. Imaging neuroinflammation—from bench to bedside. J Clin Immunol. 2014;5 https://doi.org/10.4172/2155-9899.1000226.
77. Bloomfield PS, Selvaraj S, Veronese M, et al. Microglial activity in people at ultra high risk of psychosis and in schizophrenia: an [$^{11}$C]PBR28 PET brain imaging study. Am J Psychiatry. 2016;173:44–52.
78. Setiawan E, Wilson AA, Mizrahi R, et al. Role of translocator protein density, a marker of neuroinflammation, in the brain during major depressive episodes. JAMA Psych. 2015;72:268–75.
79. Hafizi S, Tseng H-H, Rao N, et al. Imaging microglial activation in untreated first-episode psychosis: a PET study with [$^{18}$F]FEPPA. Am J Psychiatry. 2017;174(2):118–24. https://doi.org/10.1176/appi.ajp.2016.16020171.
80. Hannestad J, DellaGioia N, Gallezot J-D, et al. The neuroinflammation marker translocator protein is not elevated in individuals with mild-to-moderate depression: a [$^{11}$C]PBR28 PET study. Brain Behav Immun. 2013;33:131–8.
81. Kenk M, Selvanathan T, Rao N, et al. Imaging neuroinflammation in gray and white matter in schizophrenia: an in vivo PET study with [18F]-FEPPA. Schizophr Bull. 2015;41:85–93.
82. van der Doef TF, de Witte LD, Sutterland AL, et al. In vivo *(R)*-[$^{11}$C]PK11195 PET imaging of 18kDa translocator protein in recent onset psychosis. NPJ Schizophr. 2016;2:16031. https://doi.org/10.1038/npjschz.2016.31.
83. Golla SSV, Boellaard R, Oikonen V, et al. Quantification of [$^{18}$F]DPA-714 binding in the human brain: initial studies in healthy controls and Alzheimer's disease patients. J Cereb Blood Flow Metab. 2015;35:766–72.
84. Suridjan I, Pollock BG, Verhoeff NPLG, et al. In-vivo imaging of grey and white matter neuroinflammation in Alzheimer's disease: a positron emission tomography study with a novel radioligand, [$^{18}$F]-FEPPA. Mol Psychiatry. 2015;20:1579–87.
85. Varrone A, Oikonen V, Forsberg A, et al. Positron emission tomography imaging of the 18-kDa translocator protein (TSPO) with [$^{18}$F]FEMPA in Alzheimer's disease patients and control subjects. Eur J Nucl Med Mol Imaging. 2015;42:438–46.
86. Yokokura M, Terada T, Bunai T, et al. Depiction of microglial activation in aging and dementia: positron emission tomography with [$^{11}$C]DPA713 versus [$^{11}$C]*(R)*PK11195. J Cereb Blood Flow Metab. 2017;37(3):877–89. https://doi.org/10.1177/0271678X16646788.
87. Castanon N, Luheshi G, Laye S. Role of neuroinflammation in the emotional and cognitive alterations displayed by animal models of obesity. Front Neurosci. 2015;9:229. https://doi.org/10.3389/fnins.2015.00229.
88. Chiu CC, Liao YE, Yang LY, et al. Neuroinflammation in animal models of traumatic brain injury. J Neurosci Methods. 2016;272:38–49. https://doi.org/10.1016/j.jneumeth.2016.06.018.
89. Nazem A, Sankowski R, Bacher M, et al. Rodent models of neuroinflammation for Alzheimer's disease. J Neuroinflammation. 2015;12:74. https://doi.org/10.1186/s12974-015-0291-y.
90. Biesmans S, Acton PD, Cotto C, et al. Effect of stress and peripheral immune activation on astrocyte activation in transgenic bioluminescent Gfap-luc mice. Glia. 2015;63:1126–37.
91. Briard E, Zoghbi SS, Simeon FG, et al. Single-step high yield radiosynthesis and evaluation of a sensitive $^{18}$F-labeled ligand for imaging brain peripheral benzodiazepine receptors with PET. J Med Chem. 2009;52:688–99.
92. Buck JR, McKinley ET, Hight MR, et al. Quantitative, preclinical PET imaging of TSPO expression in glioma using [$^{18}$F]PBR06. J Nucl Med. 2011;52:107–14.

93. Fujimaru Y, Zoghbi SS, Simeon FG, et al. Quantification of translocator protein (18 kDa) in the human brain with PET and a novel radioligand (18)F-PBR06. J Nucl Med. 2009;50:1047–53.
94. Iamaizumi M, Briard E, Zoghbi SS, et al. Kinetic evaluation in nonhuman primates of two new PET ligands for peripheral benzodiazepine receptors in brain. Synapse. 2007;61:595–605.
95. Lartey FM, Ahn G-O, Shen B, et al. PET imaging of stroke-induced neuroinflammation in mice using [$^{18}$F]PBR06. Mol Imaging Biol. 2014;16:109–17.
96. Dedeurwaerdere S, Callaghan PD, Pham T, et al. PET imaging of brain inflammation during early epileptogenesis in a rat model of temporal lobe epilepsy. EJNMMI Res. 2012;2:60. https://doi.org/10.1186/2191-219X-2-60.
97. Wolf OT, Dyakin V, Patel A, et al. Volumetric structural magnetic resonance imaging (MRI) of the rat hippocampus following kainic acid (KA) treatment. Brain Res. 2002;934:87–96.
98. Zhu L, Ramboz S, Hewitt D, et al. Non-invasive imaging of GFAP expression after neuronal damage in mice. Neurosci Lett. 2004;367:210–2.
99. Biesmans S, Bouwknecht JA, Ver Donck L, et al. Peripheral administration of tumor necrosis factor-alpha induces neuroinflammation and sickness but not depressive-like behavior in mice. Biomed Res Int. 2015;2015:716920. https://doi.org/10.1155/2105/716920.
100. Biesmans S, Meert TF, Bouwknecht JA, et al. Systemic immune activation leads to neuroinflammation and sickness behavior in mice. Mediat Inflamm. 2013;2013:271359. https://doi.org/10.1155/2013/271359.
101. Cordeau P Jr, Lalancette-Hebert M, Weng YC, et al. Live imaging of neuroinflammation reveals sex and estrogen effects on astrocyte response to ischemic injury. Stroke. 2008;39:935–42.
102. Luo J, Elwood F, Britschgi M, et al. Colony-stimulating factor 1 receptor (CSF1R) signaling in injured neurons facilitates protection and survival. J Exp Med. 2013;210:157–72.
103. Golden GT, Smith GG, Ferraro TN, et al. Rat strain and age differences in kainic acid induced seizures. Epilepsy Res. 1995;20:151–9.
104. Ferraro TN, Golden GT, Smith GG, et al. Mapping murine loci for seizure response to kainic acid. Mamm Genome. 1997;8:200–8.
105. McKhann GM, Wenzel HJ, Robbins CA, et al. Mouse strain differences in kainic acid sensitivity, seizure behavior, mortality, and hippocampal pathology. Neuroscience. 2003;122:551–61.
106. Smolen JS, Aletaha D, McInnes IB. Rheumatoid arthritis. Lancet. 2016;388(10055):P2023–38. https://doi.org/10.1016/S0140-6736(16)30173-8.
107. Put S, Westhovens R, Lahoutte T, et al. Molecular imaging of rheumatoid arthritis: emerging markers, tools, and techniques. Arthritis Res Ther. 2014;16:208.
108. Wang S-C, Xie Q, Lv W-F. Positron emission tomography/computed tomography imaging and rheumatoid arthritis. Int J Rheum Dis. 2014;17:248–55.
109. Wunder A, Straub RH, Gay S, et al. Molecular imaging: novel tools in visualizing rheumatoid arthritis. Rheumatology. 2005;44:1341–9.
110. Borrero CG, Mountz JM, Mountz JD. Emerging MRI methods in rheumatoid arthritis. Nat Rev Rheumatol. 2011;7:85–95.
111. Rogers JL, Tarrant T, Kim J. Nanoparticle-based diagnostic imaging of inflammation in rheumatoid disease. Curr Rheumatol Rev. 2014;10:3–10.
112. Clavel G, Marchiol-Fournigault C, Renault G, et al. Ultrasound and Doppler micro-imaging in a model of rheumatoid arthritis. Ann Rheum Dis. 2008;67:1765–72.
113. McInnes IB, Schett G. The pathogenesis of rheumatoid arthritis. N Engl J Med. 2011;365:2205–19.
114. Ferrari M, Onuoha SC, Pitzalis C. Trojan horses and guided missiles: targeted therapies in the war on arthritis. Nat Rev Rheumatol. 2015;11:328–37.
115. Asquith DL, Miller AM, McInnes IB, et al. Animal models of rheumatoid arthritis. Eur J Immunol. 2009;39:2040–4.
116. McNamee K, Williams R, Seed M. Animals models of rheumatoid arthritis: how informative are they? Eur J Pharmacol. 2015;759:278–86.
117. Vierboom MPM, Jonker M, Tak PP, et al. Preclinical models of arthritic disease in nonhuman primates. Drug Discov Today. 2007;12:327–35.

118. Kundu-Raychaudhuri S, Mitra A, Datta-Mitra A, et al. In vivo quantification of mouse autoimmune arthritis by PET/CT. Int J Rheum Dis. 2016;19:452–8.
119. Matsui T, Nakata N, Nagai S, et al. Inflammatory cytokines and hypoxia contribute to $^{18}$F-FDG uptake by cells involved in pannus formation in rheumatoid arthritis. J Nucl Med. 2009;50:920–6.
120. Cho H, Bhatti F-U-R, Yoon TW, et al. Non-invasive dual fluorescence in vivo imaging for detection of macrophage infiltration and matrix metalloproteinase (MMP) activity in inflammatory arthritic joints. Biomed Optics Express. 2016;7:1842–52.
121. Terry SYA, Koenders MI, Franssen GM, et al. Monitoring therapy response of experimental arthritis with radiolabeled tracers targeting fibroblasts, macrophages, or integrin $\alpha_v\beta_3$. J Nucl Med. 2016;57:467–72.
122. van der Geest T, Laverman P, Gerrits D, et al. Liposomal treatment of experimental arthritis can be monitored non-invasively with radiolabeled anti-fibroblast activation protein antibodies. J Nucl Med. 2017;58(1):151–5. https://doi.org/10.2967/jnumed.116.177931.
123. Mountz JM, Alavi A, Mountz JD. Emerging optical and nuclear medicine imaging methods in rheumatoid arthritis. Nat Rev Rheumatol. 2012;8:719–28.
124. Robertson R, Germanos MS, Li C, et al. Optical imaging of Cerenkov light generation from positron-emitting radiotracers. Phys Med Biol. 2009;54:N355–66.
125. Thorek DLJ, Robertson R, Bacchus WA, et al. Cerenkov imaging—a new modality for molecular imaging. Am J Nucl Med Mol Imaging. 2012;2:163–73.
126. Irmler IM, Gebhardt P, Hoffmann B, et al. $^{18}$F-fluoride positron emission tomography/computed tomography for noninvasive in vivo quantification of pathophysiological bone metabolism in experimental murine arthritis. Arthritis Res Ther. 2014;16:R155.
127. Barck KH, Lee WP, Diehl LJ, et al. Quantification of cortical bone loss and repair for therapeutic evaluation in collagen-induced arthritis, by micro-computed tomography and automated image analysis. Arthritis Rheum. 2004;50:3377–86.
128. Sevilla RS, Cruz F, Chiu C-S, et al. Development and optimization of a high-throughput micro-computed tomography imaging method incorporating a novel analysis technique to evaluate bone mineral density of arthritic joints in a rodent model of collagen induced arthritis. Bone. 2015;73:32–41.
129. Silva MD, Ruan J, Siebert E, et al. Application of surface roughness analysis on micro-computed tomography images of bone erosion: examples using a rodent model of rheumatoid arthritis. Mol Imaging. 2006;5:475–84.
130. Silva MD, Savinainen A, Kapadia R, et al. Quantitative assessment of micro-CT imaging and histopathological signatures of experimental arthritis in rats. Mol Imaging. 2004;3:312–8.
131. Abraham C, Cho JH. Inflammatory bowel disease. N Engl J Med. 2009;361:2066–78.
132. de Souza HSP, Fiocchi C. Immunopathogenesis of IBD: current state of the art. Nat Rev Gastroenterol Hepatol. 2016;13:13–27.
133. Johnson CM, Wei C, Ensor JE, et al. Meta-analysis of colorectal cancer risk factors. Cancer Causes Control. 2013;24:1207–22.
134. Triantafillidis JK, Nasioulas G, Kosmidis PA. Colorectal cancer and inflammatory bowel disease: epidemiological, risk factors, mechanisms of carcinogenesis and prevention strategies. Anticancer Res. 2009;29:2727–37.
135. Ananthakrishnan AN, Cagan A, Cai T, et al. Colonoscopy is associated with a reduced risk for colon cancer and mortality in patients with inflammatory bowel diseases. Clin Gastroenterol Hepatol. 2015;13:322–9.
136. Al-Hawary MA, Kaza RK, Platt JF. CT enterography: concepts and advances in Crohn's disease imaging. Radiol Clin N Am. 2013;51:1–16.
137. Beltzer A, Kaulisch T, Bluhmki T, et al. Evaluation of quantitative imaging biomarkers in the DSS colitis model. Mol Imaging Biol. 2016;18:697–704.
138. Gee MS, Harisinghani MG. MRI in patients with inflammatory bowel disease. J Magn Reson Imaging. 2011;33:527–34.
139. Kilcoyne A, Kaplan JL, Gee MS. Inflammatory bowel disease imaging: current practice and future directions. World J Gastroenterol. 2016;22:917–32.

140. Masselli G, Gualdi G. CT and RM enterography in evaluating small bowel diseases: when to use which modality? Abdom Imaging. 2013;38:249–59.
141. Mentzel H-J, Reinsch S, Kurzai M, et al. Magnetic resonance imaging in children and adolescents with chronic inflammatory bowel disease. World J Gastroenterol. 2014;20:1180–91.
142. Panes J, Bouhnik Y, Reinisch W, et al. Imaging techniques for assessment of inflammatory bowel disease: joint ECCO and ESGAR evidence-based consensus guidelines. J Crohn's Colitis. 2013;7:556–85.
143. Perlman SB, Hall BS, Reichelderfer M. PET/CT imaging of inflammatory bowel disease. Semin Nucl Med. 2013;43:420–6.
144. Ream JM, Dillman JR, Adler J, et al. MRI diffusion-weighted imaging (DWI) in pediatric small bowel Crohn disease: correlation with MRI findings of active bowel wall inflammation. Pediatr Radiol. 2013;43:1077–85.
145. Saverymuttu SH, Peters AM, Hodgson HJ, et al. Indium-111 autologous leucocyte scanning: comparison with radiology for imaging the colon in inflammatory bowel disease. Brit Med J. 1982;285:255–7.
146. Treglia G, Quartuccio N, Sadeghi R, et al. Diagnostic performance of fluorine-18-fluorodeoxyglucose positron emission tomography in patients with chronic inflammatory bowel disease: a systematic review and meta-analysis. J Crohn's Colitis. 2013;7:345–54.
147. Kozarek R, Chiorean M, Wallace M, editors. Endoscopy in inflammatory bowel disease. Heidelberg: Springer; 2015.
148. Burggraaf J, Kamerling IMC, Gordon PB, et al. Detection of colorectal polyps in humans using an intravenously administered fluorescent peptide targeted against c-Met. Nat Med. 2015;21:955–61.
149. Knieling F, Waldner MJ. Light and sound—emerging techniques for inflammatory bowel disease. World J Gastroenterol. 2016;22:5642–54.
150. Liu J, Dlugosz A, Neumann H. Beyond white light endoscopy: the role of optical biopsy in inflammatory bowel disease. World J Gastroenterol. 2013;19:7544–51.
151. Subramanian V, Ragunath K. Advanced endoscopic imaging: a review of commercially available technologies. Clin Gastroenterol Hepatol. 2014;12:368–76.
152. Coda S, Thillainayagam AV. State of the art in advanced endoscopic imaging for the detection and evaluation of dysplasia and early cancer of the gastrointestinal tract. Clin Exp Gastroenterol. 2014;7:133–50.
153. Kumashiro R, Konishi K, Chiba T, et al. Integrated endoscopic system based on optical imaging and hyperspectral data analysis for colorectal cancer detection. Anticancer Res. 2016;36:3925–32.
154. Pence I, Mahadevan-Jansen A. Clinical instrumentation and applications of Raman spectroscopy. Chem Soc Rev. 2016;45:1958–79.
155. Tontini GE, Rath T, Neumann H. Advanced gastrointestinal endoscopic imaging for inflammatory bowel disease. World J Gastroenterol. 2016;22:1246–59.
156. Wu Y, Briley K, Tao X. Nanoparticle-based imaging of inflammatory bowel disease. NanoBiotechnology. 2016;8:300–15.
157. Goyal N, Rana A, Ahlawat A, et al. Animal models of inflammatory bowel disease: a review. Inflammopharmacology. 2014;22:219–33.
158. Kolios G. Animal models of inflammatory bowel disease: how useful are they really? Curr Opin Gastroenterol. 2016;32:251–7.
159. Neurath MF. Animal models of inflammatory bowel diseases: illuminating the pathogenesis of colitis, ileitis and cancer. Dig Dis. 2012;30(suppl 1):91–4.
160. Valatas V, Bamias G, Kolios G. Experimental colitis models: insights into the pathogenesis of inflammatory bowel disease and translational issues. Eur J Pharmacol. 2015;759:253–64.
161. Liu T-C, Stappenbeck TS. Genetics and pathogenesis of inflammatory bowel disease. Annu Rev Pathol. 2016;11:127–48.
162. Mizoguchi A, Takeuchi T, Himuro H, et al. Genetically engineered mouse models for studying inflammatory bowel disease. J Pathol. 2015;238:205–19.

163. Valatas V, Vakas M, Kolios G. The value of experimental models of colitis in predicting the efficacy of biological therapies for inflammatory bowel diseases. Am J Physiol Gastrointest Liver Physiol. 2013;305:G763–85.
164. Ding S, Blue RE, Morgan DR, et al. Comparison of multiple enzyme activatable near-infrared fluorescent molecular probes for detection and quantification of inflammation in murine colitis models. Inflamm Bowel Dis. 2014;20:363–77.
165. Mitsunaga M, Kosaka N, Choyke PL, et al. Fluorescence endoscopic detection of murine colitis-associated colon cancer by topically applied enzymatically rapid-activatable probe. Gut. 2013;62:1179–86.
166. Bernards N, Pottier G, Theze B, et al. In vivo evaluation of inflammatory bowel disease with the aid of μPET and the translocator protein 18 kDa radioligand [$^{18}$F]DPA-714. Mol Imaging Biol. 2015;17:67–75.
167. Heijink DM, Kleibeuker JH, Nagengast WB, et al. Total abdominal $^{18}$F-FDG uptake reflects intestinal adenoma burden in *Apc* mutant mice. J Nucl Med. 2011;52:431–6.
168. Tietz O, Wuest M, Marshall A, et al. PET imaging of cyclooxygenase-2 (COX-2) in a pre-clinical colorectal cancer model. EJNMMI Res. 2016;6:37. https://doi.org/10.1186/s13550-016-0192-9.
169. Atreya R, Neumann H, Neufert C, et al. In vivo imaging using fluorescent antibodies to tumor necrosis factor predicts therapeutic response in Crohn's disease. Nat Med. 2014;20:313–8.
170. Hua S, Marks E, Schneider JJ, et al. Advances in oral nano-delivery systems for colon targeted drug delivery in inflammatory bowel disease: selective targeting to diseased versus healthy tissue. Nanomedicine. 2015;11:1117–32.
171. Karrout Y, Dubuquoy L, Piveteau C, et al. In vivo efficacy of microbiota-sensitive coatings for colon targeting: a promising tool for IBD therapy. J Control Release. 2015;197:121–30.
172. Takedatsu H, Mitsuyama K, Torimura T. Nanomedicine and drug delivery strategies for treatment of inflammatory bowel disease. World J Gastroenterol. 2015;21:11343–52.
173. Zhang S, Emann J, Zhou A, et al. An inflammation-targeting hydrogel for local drug delivery in inflammatory bowel disease. Sci Transl Med. 2015;7:300ra128. https://doi.org/10.1126/scitranslmed.aaa5657.
174. Ali H, Weigmann B, Collnot E-M, et al. Budesonide loaded PLGA nanoparticles for targeting the inflamed intestinal mucosa—pharmaceutical characterization and fluorescence imaging. Pharm Res. 2016;33:1085–92.
175. Jin M, Yu D-G, Geraldes CFGC, et al. Theranostic fibers for simultaneous imaging and drug delivery. Mol Pharm. 2016;13:2457–65.

# 8 Preclinical Multimodality Imaging and Image Fusion in Cardiovascular Disease

James T. Thackeray

## 8.1 Introduction

The rapid expansion and deployment of multimodality imaging instrumentation including PET-CT, SPECT-CT, and more recently PET/MR has resulted in a wide range of image fusion possibilities for cardiovascular imaging [1]. In general, these fusion approaches have been limited to combining anatomic imaging with physiologic imaging modalities, particularly the fusion of CT or MR images with PET or SPECT images. Indeed, the combination of nuclear myocardial perfusion imaging with CT coronary angiography has become common in clinical practice [2]. In addition, combination of multiple radiotracers targeted to distinct physiologic pathways is more frequently applied to obtain complementary information about pathophysiology of cardiovascular disease, similar to clinical application combining perfusion and viability [3]. The principle challenge to impage fusion in preclinical applications is the frequency of image acquisition using separate imaging systems with different spatial resolution, animal positioning, anesthesia connections, gating capabilities, and dataset file type. Software packages provided with individual cameras allow for image fusion between modalities but often require conversion of the image file for semiautomated or manual co-registration. The same operations may be necessary for third-party software packages [4]. Nonetheless, image fusion bears significant potential to clarify regional image analysis and maximize the capacity of molecular imaging approaches in cardiovascular disease models. Moreover, multimodality imaging probes including nanoparticles and other synthetic nanoparticle-like structures have gained prominence over the last decade, which may have significant impact in refining image analysis and fusion tools. In this chapter, strategies and techniques for image fusion between modalities and systems will be broadly discussed in the context of cardiovascular disease, with particular attention

J. T. Thackeray (✉)
Department of Nuclear Medicine, Hannover Medical School, Hannover, Germany
e-mail: Thackeray.James@mh-hannover.de



C. Kuntner-Hannes, Y. Haemisch (eds.), *Image Fusion in Preclinical Applications*, https://doi.org/10.1007/978-3-030-02973-9_8

paid to the relevant application of these techniques to optimize the complementary nature of sequential or simultaneous physiologic and anatomic imaging.

## 8.2 PET-CT

### 8.2.1 Localization of Signal

The primary challenge in cardiovascular imaging is the localization of a cardiac signal without requisite anatomic information. This challenge is particularly significant in cases where the imaging target does not encompass the complete myocardium. Accordingly, approaches to visualize myocardial space in concert with the PET image are desirable. In some cases, this may be accomplished by fusing the PET image to a fast CT scan, providing sufficient anatomic contrast to properly localize the PET signal.

The simplest approach for cardiac localization is the alignment of the heart within the boundaries of the lung and diaphragm, which can be identified using low contrast fast CT. Vascular activity can be localized by contrast-enhanced CT acquisitions, illuminating the vessel lumen to compare with tracer distribution in the aorta, carotid, or femoral arteries. Vascular CT contrast agents used in preclinical applications tend to be iodine- or metal-rich synthetic compounds and provide exquisite definition of the arterial and venous blood pool [5, 6]. Coronary vasculature represents a formidable challenge due to the dual factors of cardiac and respiratory motion. While some motion and partial volume correction algorithms have been introduced for clinical application [7], use in small animals with faster heart rates has not been evaluated. Blood pool contrast agents also have the benefit of improved soft tissue contrast, allowing for clearer definition of cardiac anatomy and left ventricle contours.

### 8.2.2 Fusion

There are several commercially available software packages dedicated to the fusion of multimodality images. Such software packages include the vendor-supplied software installed with the multimodal cameras (e.g., Siemens Inveon Research Workplace (IRW) and 3D Image Viewer), clinical or preclinical licensed image analysis software packages (e.g., PMOD, Hermes Hybrid Viewer, Syntegra, InVivoScope), or open-source image analysis tools (e.g., AMIDE, OsiriX) [4, 8]. The choice of software depends on the image file formats generated, available computing power and operating systems, the desired applications, and operator preference. For most preclinical cardiovascular applications, the vendor-provided analysis software is sufficient for basic fusion techniques. In cases with multiple fused datasets (i.e., >3 modalities or tracers) or for vascular fusion, more sophisticated software may be required. If segmental data is desired from the datasets, optimal software can provide a polar map based on contour detection, ray detection, and/or sampling point definitions (e.g., Munich Heart, FlowQuant, Segment) [9–12].

### 8.2.3 Transformation

PET images can be rigidly matched to the CT template, relying on anatomic landmarks to align the images. This process is simplified with the full body in the field of view for both PET and CT, as can be achieved in mice with a single bed position for acquisition [13, 14]. Lowering the activity threshold display allows the boundaries of the animal thoracic cavity to be visualized, which can then be easily matched to the boundaries of the animal defined by CT. The geometric coordinates of the PET image are realigned to the CT space, allowing for a fused image to be generated. Limitations in the soft tissue contrast of standard fast-sequence small animal CT can limit the effectiveness of automated or manual fusion of radionuclide images to the CT image, particularly when the field of view is limited.

In cases with low PET signal, sequential administration of additional tracers can be of benefit. For instance, subsequent injection of $^{18}$F-sodium fluoride ($^{18}$F-NaF) may be used to obtain a bone PET map that is more easily fused to the skeleton defined on CT (Fig. 8.1). If the animal positioning in the PET acquisition is not modified between administrations of the two tracers, the transformation can be cross-applied to the original image. Similar approaches can be applied using bone imaging with SPECT-CT. Khmelinskii et al. demonstrated segmentation and visual

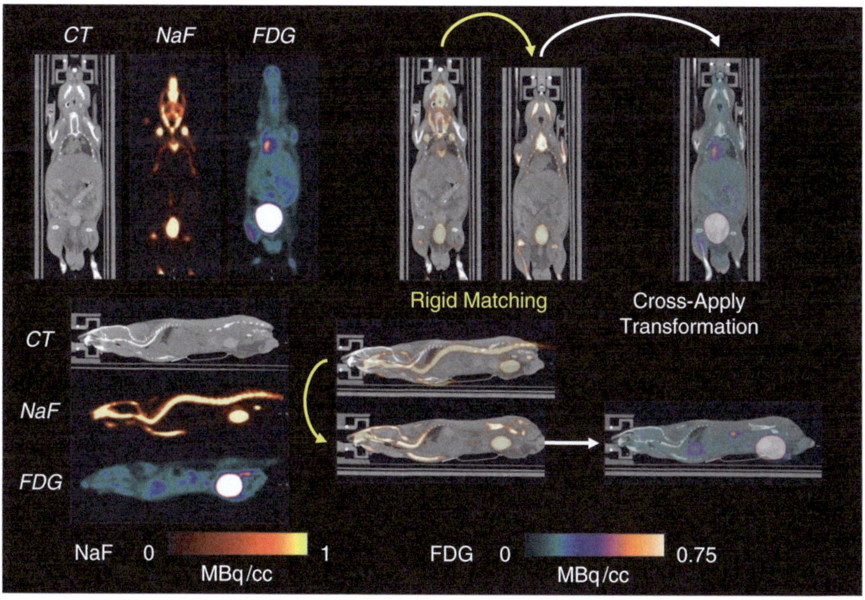


**Fig. 8.1** PET-CT image fusion. Three-step procedure involves first the fusion of the $^{18}$F-fluoride PET image to sequential fast CT scan and subsequent application of transformation coordinates to the inflammation $^{18}$F-FDG PET image. Rigid matching automatically matches the bone signal of $^{18}$F-fluoride to the skeleton defined on CT as shown in coronal and sagittal slices. The generated transformation is then cross-applied to the $^{18}$F-FDG image acquired earlier in the sequence. Image fusion carried out using Siemens Inveon Research Workplace (IRW) 3.0 software

analysis of mouse bone structures using SPECT-CT with $^{99m}$Tc-methylene diphosphonate. SPECT distribution data provided the basic skeleton, which was initially registered to the full mouse atlas and then more accurately registered by individual bone using the iterative closest point approach [15]. The articulated planar reformation algorithm then reformatted the segmented data into the mouse atlas to map space [15]. The transformation can then be cross-applied to a cardiac-specific image, enabling appropriate image fusion.

### 8.2.4 Co-registration for Multi-camera

Difficulties for moving between cameras may be compounded when different bed systems and anesthesia connections are present. Optimally, the preclinical imaging facility employs a common bed system, allowing for consistent bed alignment between cameras, though this is not always the case. Customized beds that can be moved between cameras are desirable and are available from some third-party vendors. Alternatively, animal-holding inserts that can be moved back and forth between the vendor-supplied bed systems may overcome this complication [16].

### 8.2.5 Fiducial Markers

A number of additional approaches may be employed to facilitate the fusion of images acquired using separate cameras. Fiducial markers containing CT-visualized material may be further filled with a small amount of radioactivity to visualize the marker using multiple modalities [17]. A point source placed on the scanning bed or insert that can be transferred between cameras facilitates three-dimensional co-localization of the signal between images. Such a point source may include polyethylene tubing or glass capillary tubes filled with radioactivity. These fiducial markers should be fixed to the bed and are optimally placed in multiple directions to facilitate rigid matching, without interfering with the targeted PET signal. For example, external fiducial markers optimized the positioning of the animal and bed insert between two separate dedicated scanners for sequential PET and CT acquisition. Using the point-source fiducial marker as a guide, contrast (Exia)-enhanced CT images were fused to the $^{18}$F-FDG acquisition to define the localization of the vena cava to calculate the venous input function [18].

## 8.3 SPECT-CT

Many of the same principles used for PET-CT fusion can be applied for SPECT-CT image fusion. Image transformation and the use of fiducial markers are key factors in obtaining optimal image fusion between SPECT and CT images acquired sequentially or using separate cameras (Fig. 8.2).

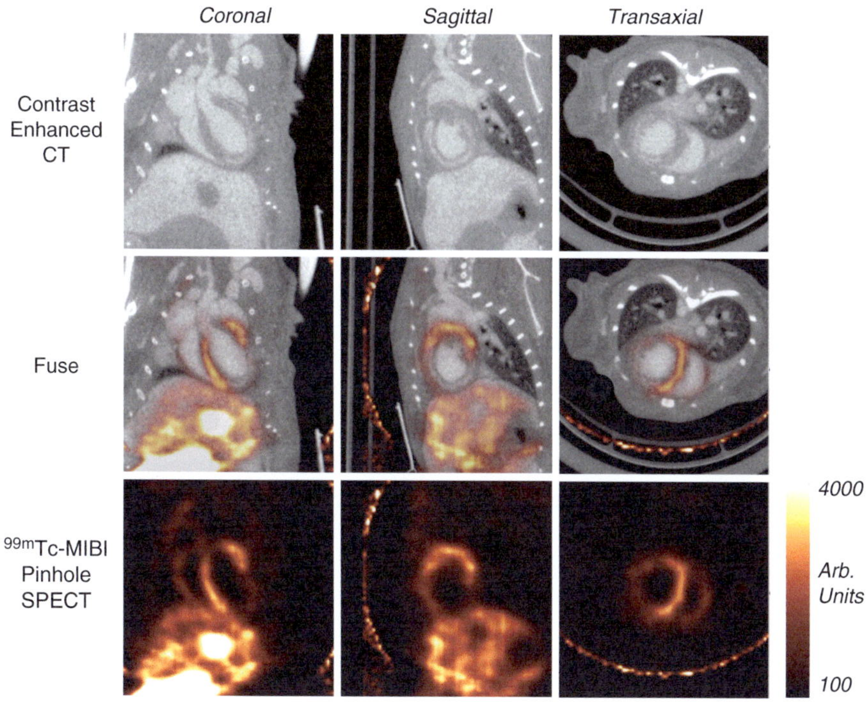


**Fig. 8.2** SPECT-CT image fusion. Contrast-enhanced CT images acquired immediately after administration of alkaline earth metal-based nanoparticle contrast agent (ExiTron nano 12,000) show high soft tissue contrast to delineate the left ventricle and ischemic region (light band) 3d after transient coronary artery occlusion. Subsequent acquisition of a perfusion SPECT study using pinhole collimation is automatically co-registered to the CT from the fused system. $^{99m}Tc$-sestamibi accumulates in the perfused myocardium, providing clear delineation of the perfusion defect in the infarct territory. Image fusion conducted using open-source AMIDE software

### 8.3.1 Localization of Signal

Fusion of SPECT images to CT may be complicated depending on the type of collimation employed for the SPECT acquisition. When using slit collimation, the positioning of individual organs is rendered more challenging due to poor image contrast and limited resolution of the image. Alternatively, improved image resolution using pinhole collimation severely limits the field of view, which may complicate precise co-registration with CT images. When using fused camera systems, the SPECT and CT images are automatically co-registered and require only minor post-acquisition manipulation to match the images to account for animal movement. The principles of alignment and co-registration described for PET above are translatable for SPECT image fusion [19].

### 8.3.2 Fusion

The image fusion within the limited space can be challenging and requires a sufficient field of view to co-localize to the CT image. The CT acquisition should exceed the boundaries of the pinhole SPECT field of view. O'Neill et al. used an antibody approach to image CD169-positive macrophages labeled with $^{99m}$Tc-pertechnetate. SPECT-CT images were analyzed using InVivoScope software and fused manually. Identification of the liver, spleen, and bone marrow was possible from fusion with CT images [20].

### 8.3.3 Transformation

Image transformation is achieved in a similar manner to PET-CT imaging, where the positioning of the SPECT image is adjusted based on rigid matching and fine-tuned with manual adjustment to fit the CT boundaries. Whether limited to the positioning of the heart within the boundaries of the lungs and diaphragm or employing higher resolution CT with or without contrast agent for soft tissue definition, rigid matching and manual adjustment perform adequately. Bone imaging agents can provide additional assistance in rigid matching [15] and may employ isotopes of different energy spectra so as not to compromise the target image. Image transformations can be saved in vendor-provided or third-party software and loaded with the image files to achieve fused images.

### 8.3.4 Fiducial Markers

Fiducial markers are applied similar to PET, with the benefit of a broader spectrum of detected activity such that different isotopes can be used for fiducial markers that are not visible in the target image reconstruction. As such, a different isotope can be used for the fiducial marker without concern for contaminating the SPECT images, regardless of the proximity to the animal in the imaging bed.

## 8.4 Multi-isotope Fusion: PET-PET, PET-SPECT, and SPECT-SPECT

Short-lived PET isotopes, different energy spectra, and complementary radiotracers open an additional category of image fusion in cardiology, namely, multi-isotope or multi-tracer fusion. Imaging protocols are designed to take advantage of these characteristics. Moreover, the emergence of next generation SPECT cameras featuring enhanced energy resolution has enabled the simultaneous acquisition and isolation of energy spectra from multiple isotopes. Clear energy resolution between thallium-201 (68 keV), technetium-99m (140 keV), iodine-123 (159 keV), and indium-111

(171 keV) allows for multi-isotope acquisition [21]. Injected activity may be administered sequentially or in a mixed bolus to minimize venous damage and to reduce variability in tracer delivery.

### 8.4.1 Regional Mismatch

In cardiology, the primary reasoning for multi-isotope studies is to isolate multiple physiological processes in a single imaging session. Typical imaging protocols would combine a molecular marker with a perfusion agent to provide definition of myocardial contours and coronary vascular territories. Combination of perfusion with glucose metabolism using $^{18}$F-FDG allows the delineation of matched defects (scar) from regions characterized by viable myocardium and hypoperfusion, termed hibernating myocardium [22]. Polar map analysis in properly fused PET-PET or PET-SPECT images facilitates regional analysis of these mismatch territories. The same principle has been applied for identifying altered myocardial sympathetic nervous integrity [23, 24] or elevated tissue inflammation following myocardial infarction or in chronic heart failure [9, 10, 25]. PET-PET fusion images rely on the short half-life of the initial imaging agent (e.g., $^{13}$N-ammonia or $^{15}$O-water for perfusion) and subsequent or delayed imaging using an agent with the longer half-life (e.g., $^{11}$C-acetate or $^{18}$F-FDG). Optimally, six half-lives should be allowed to pass between image acquisitions, but with a higher injected activity of the second tracer, the duration between scans can be shortened depending on the expected signal strength and distribution of the imaging agent. Movement of the subject between acquisitions, due to ambient settling during sustained anesthesia, injection of the second agent, or transfer between cameras in the case of PET-SPECT fusion, should be adjusted for when co-registering the acquired images. Dual isotope SPECT has the added benefit of acquiring images simultaneously, such that the images are perfectly fused without body motion. Reconstructions applying appropriate scatter windows can ideally isolate each isotope signal without loss of counts, especially when using next generation cameras equipped with cadmium-zinc-telluride (CZT) detectors.

Co-localization of an equivocal PET signal may be improved with a secondary PET image acquired with higher activity administration. For example, Thackeray et al. used $^{68}$Ga-pentixafor (~10 MBq) to image CXCR4 in activated leukocytes in the ischemic territory of myocardial infarction mouse hearts. To localize the signal, $^{18}$F-FDG (~25 MBq) was injected under isoflurane anesthesia to obtain the myocardial contours in identical position. Polar map analysis using matched sampling points facilitated the interpretation of regional distribution of the inflammation tracer with a wide positron range and low background [25]. When using $^{18}$F-FDG under ketamine-xylazine anesthesia to suppress cardiomyocyte $^{18}$F-FDG uptake, a perfusion PET scan using $^{13}$N-ammonia ~30 min prior to the administration of $^{18}$F-FDG allowed for definition of myocardial contours and subsequent fusion of perfusion and inflammation signal [9]. The relative signal intensities of the perfusion or cardiomyocyte PET signal being at least an order of magnitude higher than the inflammatory cell PET signal, there is limited cross contamination of the signal. The benefit of precise image fusion facilitates placement of regions of interest for the molecular radioligand image.

### 8.4.2 Reporter-Perfusion

These approaches have particular benefit for tracking of labeled autologous or exogenous cells for inflammation imaging or assessing experimental cell therapy. Direct cell labeling may be achieved by $^{18}$F-FDG, $^{111}$In-oxine, or $^{99m}$Tc-HMPAO, allowing cell tracking with PET or SPECT. Alternatively, cells may be genetically modified to express a reporter gene that can be imaged using targeted agents. The most commonly used reporter gene constructs are the sodium iodide symporter (NIS) and the herpes simplex virus thymidine kinase (HSV-tk), imaged using $^{99m}$Tc-pertechnetate/$^{123}$I-NaI or $^{18}$F-FHBG/$^{18}$F-FHPG, respectively [26]. The former allows for multi-isotope SPECT imaging using either $^{201}$Tl or $^{99m}$Tc-based perfusion agents in combination with $^{123}$I-NaI and, when combined with CZT detectors, may enable clear regional distribution of reporter gene-expressing cells to be realized. Appropriate co-localization of the HSV-tk-targeted PET images may be achieved in combination with $^{13}$N-ammonia to define the myocardial contours and bears potential for clear spatial orientation of signal.

Dual-energy imaging was performed in inflammation-associated amyloid in transgenic mice using a multi-isotope SPECT-CT platform. $^{125}$I-labeled serum amyloid P component (SAP) was injected with the synthetic polybasic peptide $^{99m}$Tc-p5+14 and acquired simultaneously in dual-energy SPECT, with a subsequent acquisition of CT for anatomic co-registration [27]. Imaging defined higher splenic localization of SAP versus the polybasic peptide, whereas cardiac accumulation of the synthetic peptide was higher, suggesting the potential for imaging amyloidosis with limited background [27].

Similarly, a herpes simplex virus thymidine kinase (HSV-tk) reporter gene tracked transplanted mesenchymal stromal cells using $^{18}$F-FEAU over a 5-month period. Serial PET assessed focal intramyocardial cell injection sites over time, without clear delineation of the myocardial contours. Axial CT images of the thorax effectively defined the myocardial location, though precise regional information could not be noninvasively determined [28].

Comparable approaches have been applied clinically to monitor localization of indium-111-labeled white blood cells in the setting of myocarditis [29]. CZT SPECT imaging using $^{111}$In-leukocytes with $^{99m}$Tc-sestamibi allowed for improved reader confidence and clearer definition of the myocardial contours and valve plane for diagnosis of localized inflammation [29]. The fundamental infrastructure for preclinical SPECT using a CZT-equipped camera allows the same approach but may be limited to sequential imaging in the absence of CZT detectors and their inherent energy resolution [30].

## 8.5 PET-MR-CT

PET-MR fusion imaging presents unique challenges with regard to complex acquisition protocols and the potential interference of PET instrumentation with MR acquisition. The superior soft tissue contrast afforded by MR is attractive for cardiovascular applications, as functional measurements are simplified without the requirement of contrast injection as for CT. However, unlike the fused PET-CT and SPECT-CT cameras, MR is frequently operated as a stand-alone system and generates a range of image files in different orientations which present challenges to fusion operations. Simultaneous acquisition of PET and MR data allows for optimal image fusion with limited subject movement and enables accurate and directly comparable gating of the acquired images. Nevertheless, concerns about interference with image quality have persisted.

The feasibility of PET-MR image acquisition and fusion was assessed using a standard 7T small animal MR scanner with a custom-built PET insert using lutetium oxyorthosilicate (LSO) crystals and avalanche photodiodes. In mice following myocardial infarction, $^{18}$F-FDG PET was performed in the center field of view with simultaneous acquisition of serial MRI sequences. Differences in the geometric coordinates between the cameras necessitated user interaction to manually align images. Initial evaluation using rod phantoms provided the basic transformation operations, but exact registration was required for subsequent animal image fusion. AMIDE software was employed for precise co-registration of images. While simultaneous acquisition of PET and MR measurements was feasible, the PET insert resulted in lower image resolution and sensitivity [31].

More recently, Weissler et al. reported a thorough investigation of PET and MR performance in a small animal MR camera equipped with a PET insert. Initial tests revealed a reduction of PET sensitivity inside a limited radius, with increased scatter induced by RF coils [32]. Simultaneous operation of PET and MR was initially tested using a hot-rod phantom with six hot-rod arrays with diameters and gaps of 0.8–2.0 mm filled with $^{18}$F-FDG. No difference in the PET or MR resolution was detected for individual or simultaneous acquisition. Further testing with a multinuclei phantom with a 1H/19F coil assessed the sensitivity of MR for soft tissue contrast and allowed differentiation between air, water, isopropyl alcohol, perfluoro-15-crown-5-ether, and olive oil, without loss of PET sensitivity. In vivo testing using custom-designed electrocardiogram electrodes demonstrated successful simultaneous gating of PET and MR images in eight bins. Gated PET images showed good image uniformity and excellent spatial and temporal alignment. By contrast, sequential acquisition using different camera systems requires independent gating, which results in less precise fusion in individual bins. The PET/RF Hyperion IID insert using digital silicon photomultiplier technology exhibited capability for simultaneous PET and MR image acquisition with minimal disturbance of the magnetic field and PET sensitivity [32]. The practicality of such a system, particularly considering differences in the required acquisition time for PET and MR, has not yet been established, but a clear benefit for image fusion independent of third-party software and fiducial markers offers opportunities for improvement of image analysis, gating, and reconstruction.

Nieman et al. implemented a simple modification to 3D gradient echo sequence to include a short delay between the start of the readout diphase gradient pulse and the two phase-encode pulses for imaging cardiac development in utero. Serially acquired embryo images were registered to adjust the position over the course of the scan, using freeware from the Montreal Neurological Institute. A manually drawn mask defined the embryo heart, which was then transformed for each phase of the cardiac cycle [33].

The accuracy of multimodality image fusion was previously assessed using a motion enabled targeting phantom. The phantom was molded from a human heart with comparable mechanical properties and a detachable targeting slab insert at the apex which can be customized for multiple targets. Image co-registration between MR, ultrasound, and electromagnetic signal was assessed using the phantom to provide the framework for future (pre)clinical applications [34]. Taken together, these approaches demonstrate the power of nuclear-MR image fusion approaches to combine the high spatial resolution of MR with the physiologic molecular specificity of PET or SPECT imaging.

## 8.6 Applications

### 8.6.1 Signal Localization, ROI Guidance

The primary functionality of image fusion in cardiovascular preclinical imaging is the localization of an equivocal radiotracer signal in anatomic space. Fusion of nuclear images with CT or MR can be used to guide and accurately define the positioning of appropriate regions of interest. For the myocardium, focused regional activity can be assigned to the ventricular wall based on the identification of the ventricle cavity or surrounding soft tissue. The definition of anatomy takes on higher importance for imaging of vasculature, where contrast-enhanced CT or MR images facilitate visualization of the vessel lumen. The aortic arch can be defined in this manner and, in some cases, can then be segmented to obtain planes in ascending, transverse, and descending aorta that can be compared to tissue specimens [35, 36]. For aortic aneurysm, the location of the vascular injury can be identified from the anatomic image, and the region of interest positioned independent of the radiotracer activity [37]. In these cases, fused images and guided ROI placement minimize potential bias in the image analysis and increase reader confidence in the anatomic structures.

Acquisition of $^{18}$F-FDG images followed by short CT x-ray acquisitions for attenuation correction and image co-localization was performed in C57Bl/6 mice with myocardial infarction and ischemic heart failure. CT successfully guided ROI placement in $^{18}$F-FDG images, despite reduction of glucose uptake signal by ketamine-xylazine anesthesia or conscious tracer uptake [38]. Rigid matching, manual fine-tuning, and system-matched CT images were also employed in order to accurately overlay PET images with SPECT images acquired using separate camera systems. $^{11}$C-methionine images of macrophages in the ischemic territory of mice after myocardial infarction were fused to subsequently acquired $^{99m}$Tc-sestamibi perfusion images (Fig. 8.3). PET images were first co-registered to a fast CT scan acquired on the PET-CT system, and SPECT images were co-registered to an independent fast CT scan acquired using the SPECT-CT system. The two CT images were then fused using rigid matching, and the transformation coordinates were cross-applied to the PET and SPECT images to obtain an accurate fusion of the images [10]. Perfusion images acquired using the PET scanner with $^{13}$N-ammonia

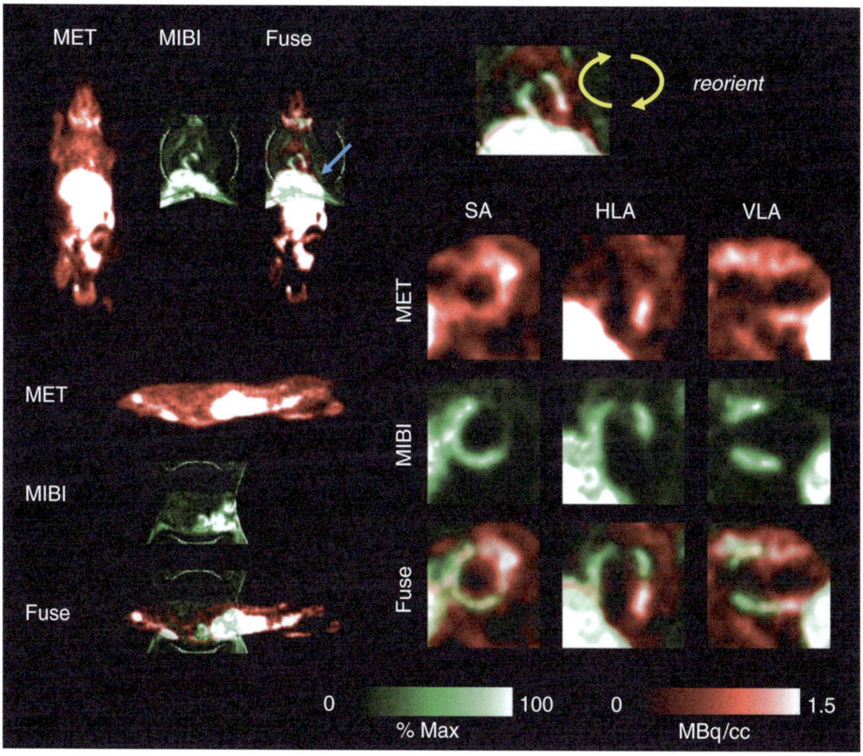


**Fig. 8.3** PET-SPECT image fusion. Fusion of PET image acquired at 30 min after injection of $^{11}$C-methionine (MET) with sequentially acquired pinhole SPECT perfusion image at 30 min after injection of $^{99m}$Tc-sestamibi (MIBI) in a mouse 3d after permanent occlusion of the left coronary artery. Pinhole SPECT field of view is restricted compared to whole body PET (left in coronal (upper) and sagittal view (lower)). Liver activity (blue arrow) is used to fine-tune the co-registration of the images from separate camera systems. The fused images are then reoriented to the cardiac axis and analyzed in matched and fused short axis (SA), horizontal long axis (HLA), and vertical long axis (VLA) slices (right). $^{11}$C-methionine accumulates in activated macrophages localized to the perfusion defect. Note the typical circular artifact around the SPECT field of view, resulting from scatter generated by pinhole collimation and the limited field of view, which does not interfere with accurate image co-registration. Images generated and co-registered using open-source AMIDE software. Modified from Thackeray et al. *Theranostics.* 2016;6:1768–79

may also provide accurate definition of the myocardial contours for co-localization of the inflammation signal [9] but requires an on-site cyclotron due to the short half-life of nitrogen-13. In the absence of perfusion images, co-registered CT can also be utilized to assign the suppressed myocardial $^{18}$F-FDG signal to the cardiac region [39], though the assignment of regions of interest using this method is less precise.

### 8.6.2 Polar Map Analysis

Multi-isotope or multi-tracer fused images lend themselves to the analysis of regional left ventricular tracer distribution, particularly when the images can be acquired simultaneously or sequentially in the same imaging session. Complementary information on myocardial perfusion, metabolism, and other molecular imaging targets enables exquisite spatial analysis of tracer distribution in the left ventricle. Dedicated polar map analysis software has been developed by several working groups and companies, often providing for interactive image co-registration and side-by-side analysis (Fig. 8.4). PET-PET, SPECT-SPECT, and PET-SPECT polar map fusion have wide-ranging applications.

**Perfusion-Metabolism**. Concurrent assessment of myocardial perfusion with regional changes in glucose or fatty acid metabolism allows for assessment of heart failure progression and monitoring of metabolism-targeted therapies. In ischemic heart failure, myocardial scar or area at risk may be defined by either PET or cardiac MR, which can allow further differentiation of the substrate. Elevated metabolism within the perfusion defect indicates repetitive stunning and/or hibernating myocardium, which bears clinical significance for revascularization [40]. In small animals, definition of altered metabolism is complicated by continuous anesthesia which affects glucose uptake by cardiomyocytes [41]. Nevertheless, combined analysis of perfusion and metabolism is feasible and provides complementary data in assessing cardiac health.

**Perfusion-Innervation**. Clinical and preclinical studies have emphasized the importance of myocardial sympathetic innervation in the progression of heart failure, sudden cardiac death, and particularly lethal ventricular arrhythmias [42, 43]. Combination of a perfusion agent ($^{13}$N-ammonia) with a sympathetic nerve-targeted tracer such as $^{11}$C-*meta*-hydroxyephedrine or $^{11}$C-epinephrine allows the definition of not only the perfusion defect, but also denervated or dysinnervated myocardium [44]. Preclinical imaging studies in pigs have confirmed the larger area of sympathetic innervation defect compared to the perfusion defect, employing a sequential imaging protocol of $^{13}$N-ammonia followed by $^{11}$C-*meta*-hydroxyephedrine or $^{11}$C-epinephrine [45]. Polar map analysis facilitates the comparison between the imaging agents and requires accurate fusion of the sequentially acquired PET images, facilitated by co-registered CT.

**Perfusion-Inflammation**. Localized inflammation within the infarct territory may be accurately defined by combined perfusion and targeted inflammatory cell imaging. Lee et al. characterized leukocyte infiltration to the infarct territory following myocardial ischemia in mouse models, using $^{18}$F-FDG imaging under

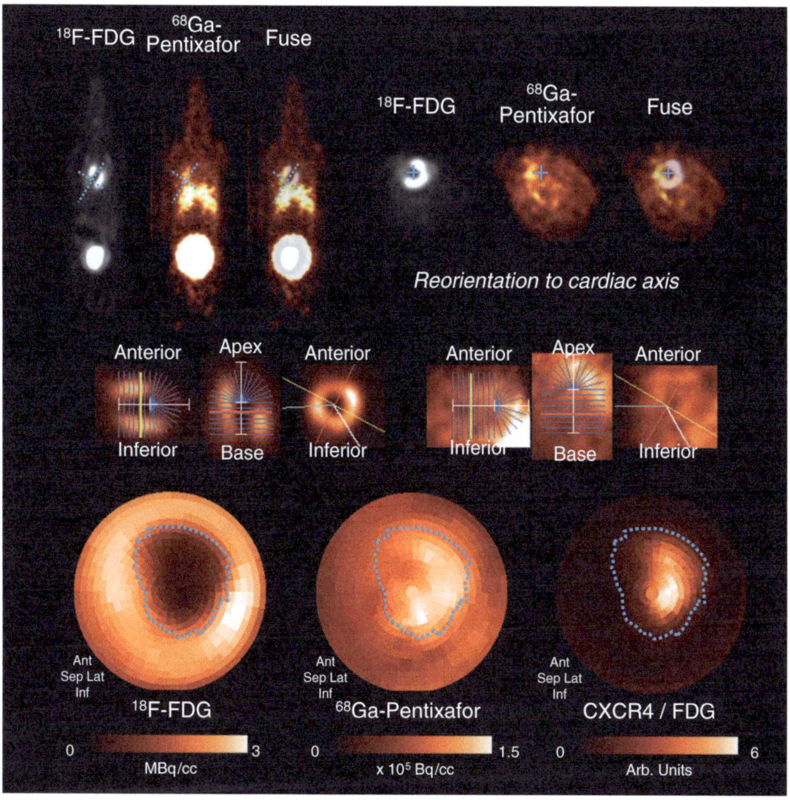


**Fig. 8.4** Polar map analysis. Sequential acquisition of PET images using first $^{68}$Ga-pentixafor targeted to CXCR4 and second $^{18}$F-FDG under isoflurane to show viable myocardium. Whole body images (top) are reoriented to the cardiac axis (blue lines), and ray detection (center) through the ventricular long axes is used to generate polar maps for analysis (bottom). The infarct region defined by $^{18}$F-FDG co-localizes to increased accumulation of $^{68}$Ga-pentixafor showing CXCR4 upregulation in the inflammatory region. Correction of the $^{68}$Ga-pentixafor image by the infarct confirms localized increase of CXCR4. Image co-registration and polar map generation performed using Munich Heart software. Modified from Thackeray et al. *JACC Cardiovasc Imaging* 2015;8:1417–26

ketamine-xylazine to suppress cardiomyocyte uptake in combination with magnetic resonance imaging. A fiducial vest was combined with a specially designed bed to allow the placement of the animal in identical positions between two scanners. The bed design facilitated no-touch transfer between scanner systems. The fiducial vest was filled with 15% iodine in water, which permitted visualization with CT and MRI, with the PET-CT fusion carried out as part of standard protocol. Fusion accuracy was tested using a $^{18}$F-filled phantom with cross-correlation function, showing a peak correlation coefficient of 0.91 over five repetitions. The fiducial vest allowed fine adjustment of the co-registration due to subtle changes in the animal position during anesthesia and transfer. Such methods were necessitated by the lack of clear myocardial contours under effective ketamine-xylazine suppression and allowed the localization of the signal to the infarct territory. Validation of the signal source was performed using flow cytometry, identifying CD11b-positive inflammatory cells as the primary contributors to the $^{18}$F-FDG signal in the infarct region [16]. Similarly, sequential imaging of macrophage-expressed mitochondrial translocator protein (TSPO) using $^{18}$F-flutriciclamide and perfusion using $^{99m}$Tc-sestamibi allowed the identification of increased macrophages infiltration in the hypoperfused region of the heart, verified by immunohistology and autoradiography [46]. Such approaches have been directly translated to clinical application [47, 48].

Image fusion is of additional benefit for analysis of multimodal imaging probes. A $^{64}$Cu-labeled nanoparticle assessed macrophage distribution in rejected cardiac allografts in mice, wherein heterotrophic abdominal grafts were imaged using PET and iodine vascular contrast-enhanced CT to localize the transplanted heart. Images were fused manually using Inveon Research Workplace and analyzed using OsiriX [49].

**Viability and Necrosis**. Definition of the infarct territory can be enhanced using both viable myocardium- and scar-avid tracers. Wang et al. acquired SPECT and CT images using a necrosis-avid agent $^{131}$I-rhein in rats after coronary artery occlusion. Images were acquired sequentially, with fusion performed using Syntegra software. Contrast-enhanced CT images provided the localization of the myocardium to focus the diffuse $^{131}$I-rhein signal in the necrotic muscle [50]. A comparable study used labeled platelets and CT to measure the gradual accumulation of $^{64}$Cu-labeled activated integrin GPIIb/IIIa sarcophagine cage complex, identifying focal platelet accumulation in the ischemic territory in a mouse model of myocardial infarction. The subsequent CT acquisition allowed definition of the focal PET signal to the ischemic territory [51].

### 8.6.3 Ventricular Remodeling

In the remodeling heart, combination of targeted molecular imaging agents with anatomic imaging can assist in ROI placement and analysis. Regional assessment may be further enhanced with perfusion analysis or angiography. Jung et al. employed a multimodality approach including $^{99m}$Tc-RP805 for SPECT imaging of matrix metalloproteinase in combination with CT angiography using iohexol (Jung et al. 2015) to define the aortic valve and assess valvular calcification. Images were co-registered automatically by the camera proprietary software (Gamma Medica X-SPECT), and a ROI was defined at the level of the aortic valve identified on the CT image. Voxels exceeding 130 Hounsfield units were defined as calcification [52]. Matrix metalloproteinase activation was proportional to aortic valve calcification but was detectable at an earlier timepoint, suggesting a capacity to predict outcome. Such approaches underscore the complementary nature of multimodality imaging and the importance of image fusion to understand the interconnection of diverse pathophysiologic processes.

### 8.6.4 Surgical Guidance

Image fusion is not limited to diagnostic techniques and has shown some potential for the direct guidance of surgical procedures and interventions. Duckett et al. applied a segmentation algorithm to overlay coronary sinus anatomy with real-time acquired fluoroscopy for the accurate placement of ventricular leads in cardiac resynchronization therapy. CT images were segmented using a 3D anatomic model using fully automated Philips EP Planner software to give models of cardiac chambers and great vessels. In a separate group of patients, segmentation was performed semiautomatically using ITK-SNAP software, with the myocardial scar mapped based on late gadolinium enhancement images. Overlay of the anatomic model on the live x-ray fluoroscopy data was performed by an experienced operator, with the heart isocentered in the x-ray field of view, then using implant electrodes for guidance of the segmented anatomic model. For cardiac MR, the right atrium was localized by a quadri-polar electrode catheter looped in the right atrium, allowing localization of the segmented anatomic model. The co-registration enabled accurate placement of the CRT leads in the left ventricle [53].

Fusion of PET data with electrophysiology maps provided insights into the contribution of cardiac sympathetic innervation to the substrate of ventricular arrhythmia. In a study of pigs after myocardial infarction, Sasano et al. fused a $^{11}$C-epinephrine polar map denoting regional defects in sympathetic neurons with an electrophysiology study showing the sites of activation of ventricular tachycardia. The sites of earliest activation of arrhythmia were co-localized to $^{11}$C-epinephrine uptake defects [54], suggesting contribution of sympathetic dysinnervation to the tachycardia substrate. This concept has been translated clinically in a subsequent study in which a $^{123}$I-MIBG SPECT scan was performed prior to ventricular ablation, with the electrophysiology map superimposed onto the 3D SPECT innervation map post hoc. The ablation sites were consistently located in denervated regions of myocardium as defined by MIBG SPECT [55]. Precise overlay of nuclear and electrophysiology data has translational potential for clinical and surgical refinement, though further characterization is warranted.

### 8.6.5 Vascular Imaging

Critically, fusion imaging can overcome limitations of defining vasculature in standalone PET or SPECT images. While such studies require the administration of contrast agent, definition of the regions of interest using the anatomic images can limit operator bias and partial volume effects on vascular image interpretation in atherosclerosis or aortic aneurysm. As a proof of concept, Fricke et al. evaluated manual registration of PET images with CT angiography from iomeprol contrast CT images. Polar maps were generated by MATLAB-based software using a PET uptake image for co-registration to the CT study and displayed in slices along the transverse, sagittal, and coronal axes. Contour lines and alpha blending allowed for dataset fusion with realignment by rotation and translation of the datasets. Manual co-registrations were compared among five operators, with an average Euclidean variation of 1.9–2.1 mm. The co-registration allows for clear visualization of culprit coronary artery with perfusion abnormality on stress or rest PET images [56].

More recently, aortic atherosclerotic plaques were targeted using $^{111}$In-tilmanocept directed at macrophage-expressed mannose receptor in a mouse model of atherosclerosis. Fusion of the CT image with the $^{111}$In-tilmanocept SPECT image enabled the accurate localization of mannose receptor within atherosclerotic plaque [57]. Other studies have applied a similar localization concept [58, 59], moving toward more standardized anatomic and physiologic assessment of plaque stability.

Clinical studies suggest the capability to translate these approaches to coronary artery disease. The multicenter EVINCI study employed hybrid imaging of CT angiography with perfusion scintigraphy to determine added value in multimodal imaging in 252 patients with stable angina and intermediate pretest likelihood of coronary artery disease. The authors emphasized the value of hybrid imaging to more accurately co-localize myocardial perfusion defects with subtending coronary arteries accounting for variability in individual coronary anatomy, with a direct impact on clinical decision-making in 20% of patients [60].

### 8.6.6 Cell Tracking

Molecular imaging techniques can be valuable in tracing the distribution and retention of transplanted cells, but single modality approaches are frequently complicated by the lack of an anatomic template to localize the cell-based signal.

Terrovitis et al. employed either SPECT-CT or PET-CT to localize transplanted NIS reporter gene-expressing cardiac-derived cardiospheres in the ischemic territory of rats after myocardial infarction. $^{99m}$Tc-pertechnetate and $^{201}$Tl were co-administered for concurrent imaging using SPECT, providing perfectly fused images of transplanted cells and perfusion, respectively. The combined SPECT images were further co-registered to a fast CT image to better localize the activity [61]. This approach was repeated using whole body PET imaging, using $^{124}$I to image NIS-positive cells, followed by a short acquisition of a higher administered dose of $^{13}$N-ammonia for perfusion. The image fusion was matched due to the identical position of the animal and further merged to a subsequently acquired CT image for anatomic localization [61]. The combination of reporter gene imaging with a perfusion or metabolic agent improves the precise localization of the transplanted cell signal.

Using a similar approach for clinical application, SPECT-MR assessed CD34+ transplanted cells in the peri-infarct zone in revascularized patients after myocardial infarction. Transplanted cells were labeled with $^{99m}$Tc-exametazime to determine the precise points of administration with subsequent perfusion imaging with $^{99m}$Tc-sestamibi to define the infarct territory. Images were fused manually using Siemens proprietary software with the border zone automatically defined by interactive thresholding of the perfusion images. Cardiac MR with late gadolinium enhancement (LGE) confirmed the infarct localization, and images were co-registered to the

$^{99m}$Tc-exametazime SPECT to appropriately localize the signal. The study identified some variability in the definition of the infarct territory and border zone between SPECT perfusion and LGE MRI, but the fusion of images provided appropriate anatomic guideposts for the location of transplanted cells [62].

Multi-isotope approaches can improve the localization of the cell-based imaging signal. Wang et al. performed dual isotope reporter gene imaging using $^{111}$In-octreotide to target somatostatin receptor type 2 (SSTR2) in macrophages and $^{99m}$Tc-pertechnetate to target sodium iodide symporter transduced in HT29 tumor cells. The different isotope energies allowed for simultaneous acquisition of SPECT images with perfect co-registration. Anatomic localization was validated using co-registered CT, with tracer accumulation comparable between NIS- and SSTR2--transduced cells [63]. The ability to precisely localize transplanted or homing cells can be invaluable to the development of therapeutic strategies to enhance engraftment, maximize recruitment, and optimize tissue repair.

## 8.7 Summary and Future Perspective

Considering the expansion of hybrid imaging technologies, it is important to take advantage of the capability to fuse images acquired using multiple modalities with complementary information. Indeed, clinical imaging frequently employs CT angiography in combination with myocardial perfusion imaging, facilitating improved visualization of culprit coronary arteries prior to revascularization [60, 64–67]. Moreover, localization of many nuclear imaging signals in the heart and vasculature lacks sufficient background signal to clearly define the signal localization without a fused guidance image, which may be acquired using CT, MR, or perfusion/viability nuclear imaging. A wealth of software packages has improved the simplicity and reproducibility of image fusion, such that three-dimensional co-registration is largely automated, requiring only minimal user interaction to optimize the overlay. Nonetheless, ideal cardiac image fusion in preclinical applications is obtained using a multi-modality integrated camera system in which the geometry, physiologic status, and positioning of the animal are identical between acquisitions. The combination of anatomic imaging with physiologic imaging can greatly improve the localization of regions of interest for quantitative image analysis, particularly when using a molecular targeted tracer that does not provide clear myocardial contours, or for vascular imaging where the target can be poorly defined without anatomic reference. It is therefore important to design the imaging protocol to take full advantage of multimodality imaging capabilities.

The precision of automated and manual image fusion can be enhanced by transferrable animal beds, fiducial markers (ideally in close proximity to the animal), dedicated transformation algorithms, and contrast agents or complementary radiotracers to define myocardial contours. Indeed, fusion of multiple isotope and diverse physiologic nuclear images bears potential to localize molecular signals more precisely. Numerous third-party softwares including open-source packages will allow continuous improvement in the precision and dissemination of these techniques.

Beyond mere signal localization, image fusion in preclinical cardiac imaging enables the concurrent evaluation of tissue substrate, metabolic function, innervation, inflammation, necrosis, and remodeling processes in myocardium and vasculature, aided by the anatomic positioning afforded by CT and/or MR imaging. Multimodality imaging probes, which combine multiple functionalities within a single targeted structure, will contribute to the expansion of image fusion techniques. In practice, cardiovascular imaging in the clinical and preclinical arena has not just approached the multimodality crossroad, but has effectively left it behind, progressing forward along a unified, fused path toward the future.

## References


1. Jaffer FA, Libby P, Weissleder R. Molecular imaging of cardiovascular disease. Circulation. 2007;116(9):1052–61.
2. Feher A, Sinusas AJ. Quantitative assessment of coronary microvascular function: dynamic single-photon emission computed tomography, positron emission tomography, ultrasound, computed tomography, and magnetic resonance imaging. Circ Cardiovasc Imaging. 2017;10(8).
3. Tragardh E, Hesse B, Knuuti J, Flotats A, Kaufmann PA, Kitsiou A, et al. Reporting nuclear cardiology: a joint position paper by the European Association of Nuclear Medicine (EANM) and the European Association of Cardiovascular Imaging (EACVI). Eur Heart J Cardiovasc Imaging. 2015;16(3):272–9.
4. Loening AM, Gambhir SS. AMIDE: a free software tool for multimodality medical image analysis. Mol Imaging. 2003;2(3):131–7.
5. Das NM, Hatsell S, Nannuru K, Huang L, Wen X, Wang L, et al. In vivo quantitative microcomputed tomographic analysis of vasculature and organs in a normal and diseased mouse model. PLoS One. 2016;11(2):e0150085.
6. Yan D, Zhang Z, Luo Q, Yang X. A novel mouse segmentation method based on dynamic contrast enhanced micro-CT images. PLoS One. 2017;12(1):e0169424.
7. Rubeaux M, Joshi NV, Dweck MR, Fletcher A, Motwani M, Thomson LE, et al. Motion correction of 18F-NaF PET for imaging coronary atherosclerotic plaques. J Nucl Med. 2016;57(1):54–9.
8. Karlsen OT, Verhagen R, Bovee WM. Parameter estimation from Rician-distributed data sets using a maximum likelihood estimator: application to T1 and perfusion measurements. Magn Reson Med. 1999;41(3):614–23.
9. Thackeray JT, Bankstahl JP, Wang Y, Korf-Klingebiel M, Walte A, Wittneben A, et al. Targeting post-infarct inflammation by PET imaging: comparison of (68)Ga-citrate and (68)Ga-DOTATATE with (18)F-FDG in a mouse model. Eur J Nucl Med Mol Imaging. 2015;42(2):317–27.
10. Thackeray JT, Bankstahl JP, Wang Y, Wollert KC, Bengel FM. Targeting amino acid metabolism for molecular imaging of inflammation early after myocardial infarction. Theranostics. 2016;6(11):1768–79.
11. Lamoureux M, Thorn S, Dumouchel T, Renaud JM, Klein R, Mason S, et al. Uniformity and repeatability of normal resting myocardial blood flow in rats using [13N]-ammonia and small animal PET. Nucl Med Commun. 2012;33(9):917–25.
12. Thackeray JT, deKemp RA, Beanlands RS, DaSilva JN. Early diabetes treatment does not prevent sympathetic dysinnervation in the streptozotocin diabetic rat heart. J Nucl Cardiol. 2014;21(4):829–41.
13. Shekhar R, Walimbe V, Raja S, Zagrodsky V, Kanvinde M, Wu G, et al. Automated 3-dimensional elastic registration of whole-body PET and CT from separate or combined scanners. J Nucl Med. 2005;46(9):1488–96.

14. Suh JW, Kwon OK, Scheinost D, Sinusas AJ, Cline GW, Papademetris X. CT-PET weighted image fusion for separately scanned whole body rat. Med Phys. 2012;39(1):533–42.
15. Khmelinskii A, Groen HC, Baiker M, de Jong M, Lelieveldt BP. Segmentation and visual analysis of whole-body mouse skeleton microSPECT. PLoS One. 2012;7(11):e48976.
16. Lee WW, Marinelli B, van der Laan AM, Sena BF, Gorbatov R, Leuschner F, et al. PET/MRI of inflammation in myocardial infarction. J Am Coll Cardiol. 2012;59(2):153–63.
17. Dogdas B, Stout D, Chatziioannou AF, Leahy RM. Digimouse: a 3D whole body mouse atlas from CT and cryosection data. Phys Med Biol. 2007;52(3):577–87.
18. Thorn SL, deKemp RA, Dumouchel T, Klein R, Renaud JM, Wells RG, et al. Repeatable noninvasive measurement of mouse myocardial glucose uptake with 18F-FDG: evaluation of tracer kinetics in a type 1 diabetes model. J Nucl Med. 2013;54(9):1637–44.
19. Matsunari I, Miyazaki Y, Kobayashi M, Nishi K, Mizutani A, Kawai K, et al. Performance evaluation of the eXplore speCZT preclinical imaging system. Ann Nucl Med. 2014;28(5):484–97.
20. O'Neill AS, Terry SY, Brown K, Meader L, Wong AM, Cooper JD, et al. Non-invasive molecular imaging of inflammatory macrophages in allograft rejection. EJNMMI Res. 2015;5(1):69.
21. Pissarek M, Meyer-Kirchrath J, Hohlfeld T, Vollmar S, Oros-Peusquens AM, Flogel U, et al. Targeting murine heart and brain: visualisation conditions for multi-pinhole SPECT with (99m)Tc- and (123)I-labelled probes. Eur J Nucl Med Mol Imaging. 2009;36(9):1495–509.
22. Di Carli MF, Asgarzadie F, Schelbert HR, Brunken RC, Rokhsar S, Maddahi J. Relation of myocardial perfusion at rest and during pharmacologic stress to the PET patterns of tissue viability in patients with severe left ventricular dysfunction. J Nucl Cardiol. 1998;5(6):558–66.
23. Luisi AJ Jr, Suzuki G, Dekemp R, Haka MS, Toorongian SA, Canty JM Jr, et al. Regional 11C-hydroxyephedrine retention in hibernating myocardium: chronic inhomogeneity of sympathetic innervation in the absence of infarction. J Nucl Med. 2005;46(8):1368–74.
24. Yukinaka M, Nomura M, Ito S, Nakaya Y. Mismatch between myocardial accumulation of 123I-MIBG and 99mTc-MIBI and late ventricular potentials in patients after myocardial infarction: association with the development of ventricular arrhythmias. Am Heart J. 1998;136(5):859–67.
25. Thackeray JT, Derlin T, Haghikia A, Napp LC, Wang Y, Ross TL, et al. Molecular Imaging of the Chemokine Receptor CXCR4 After Acute Myocardial Infarction. JACC Cardiovasc Imaging. 2015;8(12):1417–26.
26. Higuchi T, Anton M, Saraste A, Dumler K, Pelisek J, Nekolla SG, et al. Reporter gene PET for monitoring survival of transplanted endothelial progenitor cells in the rat heart after pretreatment with VEGF and atorvastatin. J Nucl Med. 2009;50(11):1881–6.
27. Martin EB, Williams A, Richey T, Stuckey A, Heidel RE, Kennel SJ, et al. Comparative evaluation of p5+14 with SAP and peptide p5 by dual-energy SPECT imaging of mice with AA amyloidosis. Sci Rep. 2016;6:22695.
28. Perin EC, Tian M, Marini FC 3rd, Silva GV, Zheng Y, Baimbridge F, et al. Imaging long-term fate of intramyocardially implanted mesenchymal stem cells in a porcine myocardial infarction model. PLoS One. 2011;6(9):e22949.
29. Caobelli F, Wollenweber T, Bavendiek U, Kuhn C, Schutze C, Geworski L, et al. Simultaneous dual-isotope solid-state detector SPECT for improved tracking of white blood cells in suspected endocarditis. Eur Heart J. 2017;38(6):436–43.
30. Thackeray JT, Korf-Klingebiel M, Wang Y, Kustikova OS, Bankstahl JP, Wollert KC, et al. Non-invasive tracking of endogenous bone marrow cell recruitment after myocardial infarction in mice (Abstract). Circulation. 2015;132:A17146.
31. Buscher K, Judenhofer MS, Kuhlmann MT, Hermann S, Wehrl HF, Schafers KP, et al. Isochronous assessment of cardiac metabolism and function in mice using hybrid PET/MRI. J Nucl Med. 2010;51(8):1277–84.
32. Weissler B, Gebhardt P, Dueppenbecker PM, Wehner J, Schug D, Lerche CW, et al. A digital preclinical PET/MRI insert and initial results. IEEE T Med Imaging. 2015;34(11):2258–70.
33. Nieman BJ, Szulc KU, Turnbull DH. Three-dimensional, in vivo MRI with self-gating and image coregistration in the mouse. Magn Reson Med. 2009;61(5):1148–57.

34. Hatt CR, Stanton D, Parthasarathy V, Jain AK, Raval AN. A method for measuring the accuracy of multi-modal image fusion system for catheter-based cardiac interventions using a novel motion enabled targeting phantom. Conf Proc IEEE Eng Med Biol Soc. 2011;2011:6260–4.
35. Jin H, Yang H, Liu H, Zhang Y, Zhang X, Rosenberg AJ, et al. A promising carbon-11-labeled sphingosine-1-phosphate receptor 1-specific PET tracer for imaging vascular injury. J Nucl Cardiol. 2017;24(2):558–70.
36. Majmudar MD, Keliher EJ, Heidt T, Leuschner F, Truelove J, Sena BF, et al. Monocyte-directed RNAi targeting CCR2 improves infarct healing in atherosclerosis-prone mice. Circulation. 2013;127(20):2038–46.
37. Golestani R, Razavian M, Nie L, Zhang J, Jung JJ, Ye Y, et al. Imaging vessel wall biology to predict outcome in abdominal aortic aneurysm. Circ Cardiovasc Imaging. 2015;8(1)
38. Hong X, Bu L, Wang Y, Xu J, Wu J, Huang Y, et al. Increases in the risk of cognitive impairment and alterations of cerebral beta-amyloid metabolism in mouse model of heart failure. PLoS One. 2013;8(5):e63829.
39. Cusso L, Vaquero JJ, Bacharach S, Desco M. Comparison of methods to reduce myocardial 18F-FDG uptake in mice: calcium channel blockers versus high-fat diets. PLoS One. 2014;9(9):e107999.
40. D'Egidio G, Nichol G, Williams KA, Guo A, Garrard L, deKemp R, et al. Increasing benefit from revascularization is associated with increasing amounts of myocardial hibernation: a substudy of the PARR-2 trial. JACC Cardiovasc Imaging. 2009;2(9):1060–8.
41. Thackeray JT, Bankstahl JP, Wang Y, Wollert KC, Bengel FM. Clinically relevant strategies for lowering cardiomyocyte glucose uptake for 18F-FDG imaging of myocardial inflammation in mice. Eur J Nucl Med Mol Imaging. 2015;42(5):771–80.
42. Thackeray JT, Bengel FM. PET imaging of the autonomic nervous system. Q J Nucl Med Mol Imaging. 2016;60(4):362–82.
43. Fallavollita JA, Heavey BM, Luisi AJ Jr, Michalek SM, Baldwa S, Mashtare TL Jr, et al. Regional myocardial sympathetic denervation predicts the risk of sudden cardiac arrest in ischemic cardiomyopathy. J Am Coll Cardiol. 2014;63(2):141–9.
44. Thackeray JT, Renaud JM, Kordos M, Klein R, Dekemp RA, Beanlands RS, et al. Test-retest repeatability of quantitative cardiac 11C-meta-hydroxyephedrine measurements in rats by small animal positron emission tomography. Nucl Med Biol. 2013;40(5):676–81.
45. Lautamaki R, Sasano T, Higuchi T, Nekolla SG, Lardo AC, Holt DP, et al. Multiparametric molecular imaging provides mechanistic insights into sympathetic innervation impairment in the viable infarct border zone. J Nucl Med. 2015;56(3):457–63.
46. Thackeray JT, Hupe HC, Wang Y, Bankstahl JP, Berding G, Ross TL, et al. Myocardial Inflammation Predicts Remodeling and Neuroinflammation After Myocardial Infarction. J Am Coll Cardiol. 2018;71(3):263–75.
47. Wollenweber T, Roentgen P, Schafer A, Schatka I, Zwadlo C, Brunkhorst T, et al. Characterizing the inflammatory tissue response to acute myocardial infarction by clinical multimodality noninvasive imaging. Circ Cardiovasc Imaging. 2014;7(5):811–8.
48. Rischpler C, Dirschinger RJ, Nekolla SG, Kossmann H, Nicolosi S, Hanus F, et al. Prospective evaluation of 18F-fluorodeoxyglucose uptake in postischemic myocardium by simultaneous positron emission tomography/magnetic resonance imaging as a prognostic marker of functional outcome. Circ Cardiovasc Imaging. 2016;9(4):e004316.
49. Ueno T, Dutta P, Keliher E, Leuschner F, Majmudar M, Marinelli B, et al. Nanoparticle PET-CT detects rejection and immunomodulation in cardiac allografts. Circ Cardiovasc Imaging. 2013;6(4):568–73.
50. Wang Q, Yang S, Jiang C, Li J, Wang C, Chen L, et al. Discovery of radioiodinated monomeric anthraquinones as a novel class of necrosis avid agents for early imaging of necrotic myocardium. Sci Rep. 2016;6:21341.
51. Ziegler M, Alt K, Paterson BM, Kanellakis P, Bobik A, Donnelly PS, et al. Highly sensitive detection of minimal cardiac ischemia using positron emission tomography imaging of activated platelets. Sci Rep. 2016;6:38161.

52. Jung JJ, Razavian M, Challa AA, Nie L, Golestani R, Zhang J, et al. Multimodality and molecular imaging of matrix metalloproteinase activation in calcific aortic valve disease. J Nucl Med. 2015;56(6):933–8.
53. Duckett SG, Ginks MR, Knowles BR, Ma Y, Shetty A, Bostock J, et al. Advanced image fusion to overlay coronary sinus anatomy with real-time fluoroscopy to facilitate left ventricular lead implantation in CRT. Pacing Clin Electrophysiol. 2011;34(2):226–34.
54. Sasano T, Abraham MR, Chang KC, Ashikaga H, Mills KJ, Holt DP, et al. Abnormal sympathetic innervation of viable myocardium and the substrate of ventricular tachycardia after myocardial infarction. J Am Coll Cardiol. 2008;51(23):2266–75.
55. Klein T, Abdulghani M, Smith M, Huang R, Asoglu R, Remo BF, et al. Three-dimensional 123I-meta-iodobenzylguanidine cardiac innervation maps to assess substrate and successful ablation sites for ventricular tachycardia: feasibility study for a novel paradigm of innervation imaging. Circ Arrhythm Electrophysiol. 2015;8(3):583–91.
56. Fricke H, Elsner A, Weise R, Bolte M, van den Hoff J, Burchert W, et al. Quantitative myocardial perfusion PET combined with coronary anatomy derived from CT angiography: validation of a new fusion and visualisation software. Z Med Phys. 2009;19(3):182–8.
57. Varasteh Z, Hyafil F, Anizan N, Diallo D, Aid-Launais R, Mohanta S, et al. Targeting mannose receptor expression on macrophages in atherosclerotic plaques of apolipoprotein E-knockout mice using 111In-tilmanocept. EJNMMI Res. 2017;7(1):40.
58. Luehmann HP, Detering L, Fors BP, Pressly ED, Woodard PK, Randolph GJ, et al. PET/CT imaging of chemokine receptors in inflammatory atherosclerosis using targeted nanoparticles. J Nucl Med. 2016;57(7):1124–9.
59. Hyafil F, Pelisek J, Laitinen I, Schottelius M, Mohring M, Doring Y, et al. Imaging the cytokine receptor CXCR4 in atherosclerotic plaques with the radiotracer (68)Ga-pentixafor for PET. J Nucl Med. 2017;58(3):499–506.
60. Liga R, Vontobel J, Rovai D, Marinelli M, Caselli C, Pietila M, et al. Multicentre multi-device hybrid imaging study of coronary artery disease: results from the EValuation of INtegrated Cardiac Imaging for the Detection and Characterization of Ischaemic Heart Disease (EVINCI) hybrid imaging population. Eur Heart J Cardiovasc Imaging. 2016;17(9):951–60.
61. Terrovitis J, Kwok KF, Lautamaki R, Engles JM, Barth AS, Kizana E, et al. Ectopic expression of the sodium-iodide symporter enables imaging of transplanted cardiac stem cells in vivo by single-photon emission computed tomography or positron emission tomography. J Am Coll Cardiol. 2008;52(20):1652–60.
62. Musialek P, Tekieli L, Kostkiewicz M, Miszalski-Jamka T, Klimeczek P, Mazur W, et al. Infarct size determines myocardial uptake of CD34+ cells in the peri-infarct zone: results from a study of (99m)Tc-extametazime-labeled cell visualization integrated with cardiac magnetic resonance infarct imaging. Circ Cardiovasc Imaging. 2013;6(2):320–8.
63. Wang J, Arulanandam R, Wassenaar R, Falls T, Petryk J, Paget J, et al. Enhancing expression of functional human sodium iodide symporter and somatostatin receptor in recombinant oncolytic vaccinia virus for in vivo imaging of tumors. J Nucl Med. 2017;58(2):221–7.
64. de Knegt MC, Fuchs A, Weeke P, Mogelvang R, Hassager C, Kofoed KF. Optimisation of coronary vascular territorial 3D echocardiographic strain imaging using computed tomography: a feasibility study using image fusion. Int J Cardiovasc Imaging. 2016;32(12):1715–23.
65. Gaemperli O, Saraste A, Knuuti J. Cardiac hybrid imaging. Eur Heart J Cardiovasc Imaging. 2012;13(1):51–60.
66. Grant EK, Faranesh AZ, Cross RR, Olivieri LJ, Hamann KS, O'Brien KJ, et al. Image fusion guided device closure of left ventricle to right atrium Shunt. Circulation. 2015;132(14):1366–7.
67. Plank F, Mueller S, Uprimny C, Hangler H, Feuchtner G. Detection of bioprosthetic valve infection by image fusion of (18)fluorodeoxyglucose-positron emission tomography and computed tomography. Interact Cardiovasc Thorac Surg. 2012;14(3):364–6.

# Dual-Energy SPECT Imaging with Contrast-Enhanced CT: A Case Study

9

Emily B. Martin, Alan Stuckey, Stephen J. Kennel, and Jonathan S. Wall

## 9.1 Introduction to Imaging Modalities

### 9.1.1 SPECT Imaging

Single-photon emission computed tomography (SPECT) is the most common molecular imaging modality in the clinic and requires the use of an injectable tracer radiolabeled with isotopes that emit higher-energy (>140 keV) gamma ray photons. The most commonly used SPECT radionuclides in the clinic are iodine-123 ($^{123}$I; 159 keV gamma) with a half-life of 13 h and technetium-99m ($^{99m}$Tc; 140 keV gamma) with a half-life of 6.1 h. In preclinical studies, however, the low-energy emitter iodine-125 ($^{125}$I; ~30 keV) is a suitable alternative because it has a much longer half-life (59 days) allowing for extended observations and, despite the low energy of the emitted photons, is suitable for imaging small subjects such as mice. SPECT imaging requires the detection of emitted photons that are collimated prior to interaction with the detectors that, in small animal imaging systems, either rotate around or surround the subject to generate a 3D image [1, 2]. Several versions of these scanners are commercially available. Imaging systems with more than one detector have increased sensitivity. SPECT imaging offers several advantages when compared to positron emission tomography (PET). The most notable advantages are the ability to use a range of radionuclides with longer physical half-lives than those routinely used in PET protocols, which facilitate correlative preclinical studies such as high-resolution microautoradiography, and the ability to detect multiple radiotracers with different photon emission energies (such as $^{125}$I and $^{99m}$Tc) simultaneously *in vivo* [3].

E. B. Martin · A. Stuckey · S. J. Kennel · J. S. Wall (✉)
Department of Medicine, University of Tennessee Medical Center, Knoxville, TN, USA
e-mail: jwall@utmck.edu





C. Kuntner-Hannes, Y. Haemisch (eds.), *Image Fusion in Preclinical Applications*, https://doi.org/10.1007/978-3-030-02973-9_9

### 9.1.2 CT Imaging

X-ray computed tomography (CT) is a high-resolution anatomic imaging modality capable of visualizing structures based on their relative attenuation of externally applied X-rays [4]. Spatial resolution is dependent upon many factors; however, for most preclinical imaging systems, 50–150 μm resolution is used for imaging studies. Three-dimensional images are generated from multiple X-ray projections that are acquired as the X-ray source and detector rotate around the subject. As the X-rays travel through diverse types of material (e.g., fat, tissue, bone, air), they become variably attenuated, which generates a high-resolution attenuation map or CT image of the subject in which many anatomic structures can be readily visualized. While CT provides much more information than X-ray radiography, it exposes individuals to more radiation [5], but to further address this concern, faster multidetector CT scanners with the automatic exposure control have been implemented into many clinical and preclinical imaging platforms [5, 6]. Dual modality imaging has become the norm, and CT is frequently coupled with SPECT and PET in a single imaging system.

One limitation of small animal CT imaging is that organ boundaries, especially within the abdominal cavity of mice, can be difficult to discern because the X-ray attenuation is similar for those organs. To improve anatomical visualization, electron dense X-ray contrast agents, which appear white in CT images, can be administered. Contrast agents used in the clinic are concentrates of highly attenuating elements commonly administered intravascularly (typically an iodine-based compound, e.g., Iohexol™) or orally (commonly a barium-based compound) in the clinical setting which can aid in the detection of vascular or intestinal abnormalities, respectively. In the preclinical setting, when imaging mice or other small animals with little visceral fat (since fat can provide a natural contrast to discern organs [7]), reverse contrast techniques can be used, wherein intraperitoneal (i.p.) injection of contrast agents, such as Iohexol, can assist in abdominal organ delineation [8]. This technique can dramatically improve interpretation of the CT image and allow precise identification of abdominal organs that can be "segmented" in the image such that volumetric analyses can be performed and whole organ volumes, or tumor volumes, can be established (Fig. 9.1). Increased contrast between the soft abdominal soft tissues may be achieved by acquiring CT data with decreased X-ray voltage; however, image integrity may be compromised using this approach with little increase in information. In the clinic, dual-energy (spectral) CT can now be used, in conjunction with or without IV contrast agents, to generate image data based on the tissues attenuation characteristics [9–11].

### 9.1.3 Dual Modality Imaging

In many instances, CT imaging can be informative as a singular modality; however, as noted above, CT has been used in conjunction with SPECT (the focus of this

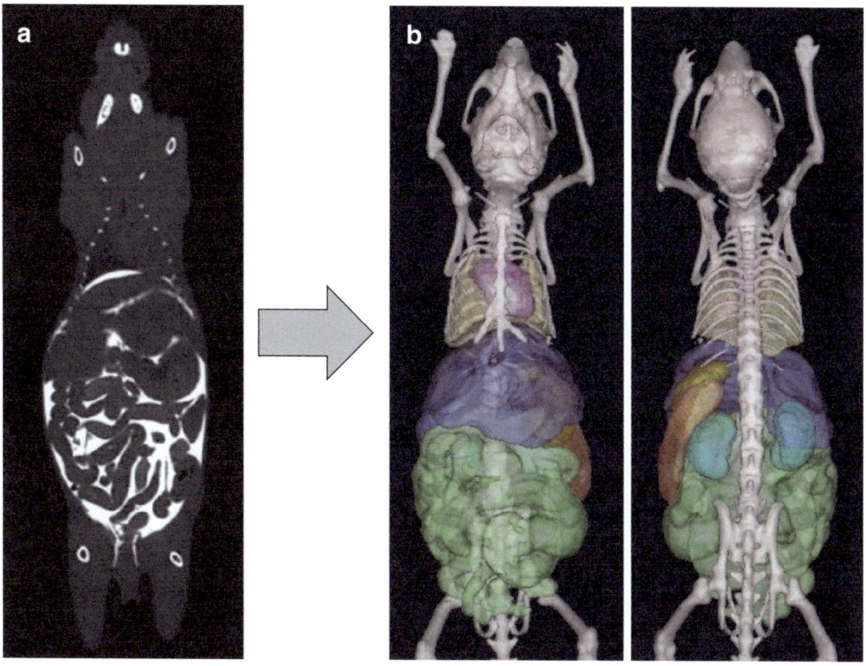


**Fig. 9.1** Complete segmentation of mouse abdominothoracic organs using contrast-enhanced CT images. (**a**) Representative 2D sagittal image of a contrast-enhanced mouse CT. The contrast enables organ delineation and facilitates segmentation. (**b**) A 3D rendering which shows ventral (left) and dorsal (right) views of fully segmented lungs, heart, liver, spleen, kidneys, stomach, and intestines

chapter) and PET to provide enhanced capabilities including providing data for scatter correction of the PET and SPECT images as well as, most importantly, providing an anatomic frame of reference to enhance interpretation of the PET and SPECT data [12]. As stated by Mariani et al., "the main advantages of SPECT/CT are represented by better attenuation correction, increased specificity and accurate depiction of the localization of disease" [13]. Additionally, SPECT/CT imaging improves visualization of functional and metabolic information that is provided with nuclear medicine tests which monitor molecular changes that may not be observed with anatomical tests alone [13–16]. SPECT/CT imaging platforms are utilized extensively in the preclinical setting for research and drug development purposes and have proven to be a valuable imaging tool for translational research most commonly pertaining to oncology and neuroscience [14]. Moreover, as we demonstrate in one example below, contrast-enhanced CT imaging may be used in conjunction with SPECT imaging to better estimate appropriate anatomical placement of regions of interest when performing radiotracer biodistribution analysis from SPECT images (see Fig. 9.2).

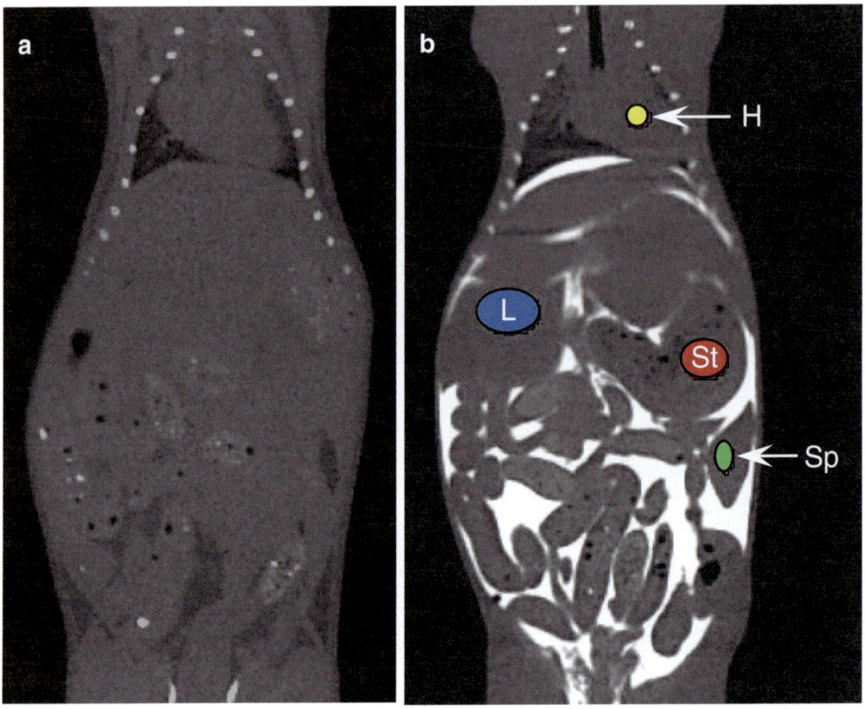


**Fig. 9.2** Contrast-enhanced CT imaging aids in placement of ROI. (**a**) Mouse CT image without contrast limits the ability of delineate organ boundaries. (**b**) Mouse CT image enhanced with i.p. contrast agent visualizes organ boundaries and facilitates proper placement of ROI in the heart (H), liver (L), stomach (St), and spleen (Sp). The kidneys were also assessed but are not visible in this CT slice

### 9.1.4 Dual-Energy SPECT/CT Imaging

A major advantage of SPECT imaging over PET imaging is the ability to assess the comparative efficacy of two radiotracers simultaneously within a single subject by employing dual-energy imaging. This is achieved by exploiting the energy properties of different radioisotopes such that tracers are radiolabeled with high- or low-energy gamma photon-emitting nuclides or isotopes (i.e., $^{123}I$ = 159 keV paired with $^{125}I$ = 20–35 keV [in the preclinical setting]). This capability allows direct comparison of two imaging agents under the same physiological conditions within the same individual; for example, the gold standard agent can be compared to a novel reagent. It also permits simultaneous visualization of different physical processes, such as the measurement of blood flow metabolic rates or protein binding, when appropriate radiotracers are available [17–23]. The ability to detect and acquire multiple measurements within an individual subject is particularly useful in murine models of complex disease where inherent dynamic biological variability perpetually exists. Using dual-energy SPECT imaging circumvents the questionable reproducibility of in vivo conditions that may arise when large independent groups of animals are used to compare the biodistribution of radiotracers in vivo. In this way, it also reduces the number of animals that must be used to achieve statistically significant results. Despite the simultaneous injection of the dual energy-emitting nuclides, the system we use acquires the high- and low-energy data sequentially. In the original MicroCAT II SPECT/CT system, simultaneous acquisition of high- and low-energy photons was possible; however, in this instance, the system used independent detectors for high- and low-energy emission. Given these approaches to data acquisition, parameters (such as detector voltage) can be optimized for each nuclide. In our system, the major concern is spill-down of high-energy photons into the low-energy window, but this is only a major concern when using $^{123}$I with $^{125}$I, and we have demonstrated that appropriate corrections can be made [24]. The following case study utilizing preclinical dual-energy SPECT/CT imaging of independent peptide radiotracers in mice serves as a workable example that can be adapted to numerous investigational studies.

## 9.2 Case Study: The Use of Dual-Energy SPECT/CT to Compare the Biodistribution of Two Peptide Radiotracers in Mice with Systemic Amyloidosis

### 9.2.1 Introduction to the Complex Disease Model: Murine Systemic AA Amyloidosis

Systemic amyloidosis refers to a class of diseases characterized by the overproduction of misfolded proteins which ultimately leads to the formation and deposition of proteinaceous fibrils, known as amyloid, in numerous and variable tissues and organs. These deposits jeopardize the structural integrity and function of affected organs, altering physiological and catabolic processes depending on the amount of amyloid in each organ. The insidious accumulation of amyloid, notably in the heart and kidney, can ultimately lead to death [25]. Therapeutic advancements have been made in recent years with various immunotherapies under evaluation in clinical trial, but the key to enhancing patient survival is early detection of amyloidosis. In Europe, radiolabeled serum amyloid P component is used for imaging abdominal amyloidosis [26]; however, despite radiotracers that can detect cardiac amyloidosis [27–30], there are no approved imaging agents available in the USA for visualizing whole-body amyloid load [18, 31]. To address this issue, we have been developing and evaluating, in an iterative fashion, a series of synthetic peptides with different bioactivities as potential agents for targeting amyloid and that, when radiolabeled, can be used as imaging agents. We have used dual-energy SPECT/CT imaging to directly compare the efficacy of our peptides in both healthy (wild type; WT) mice and those with systemic multi-organ inflammation-associated (AA) amyloidosis as a way to identify the optimal agent for translation to the clinic. Herein, we describe techniques to quantitatively compare two amyloid-binding peptides in a single subject (mouse) with amyloid disease.

### 9.2.2 Evaluation of Two Radiolabeled Peptides in Healthy Mice

The two polybasic peptide radiotracers analyzed in this study, known as $^{aqa}$p5 and $^{AQA}$p5 (radiolabeled with $^{125}$I and $^{123}$I, respectively), have the same amino acid sequence, but the N-terminal 3 amino acids of peptide $^{aqa}$p5 are D-enantiomers (denoted with lowercase letters) as opposed to the all L form, peptide $^{AQA}$p5. Incorporation of D-amino acids prevents dehalogenation of radioiodinated $^{aqa}$p5 in vivo, while $^{AQA}$p5 undergoes both renal and hepatic catabolism resulting in dehalogenation and thus redistribution of the radioiodide [32]. The goal of this study is to compare, by dual-energy SPECT/CT imaging, the distribution of radiolabeled peptide (and free radioiodide) in the same mouse. When $^{125}$I-$^{aqa}$p5 and $^{123}$I-$^{AQA}$p5 were administered intravenously (in the lateral tail vein) to healthy WT mice, the biodistribution of the two radiotracers was discrete and readily visualized (Fig. 9.3).

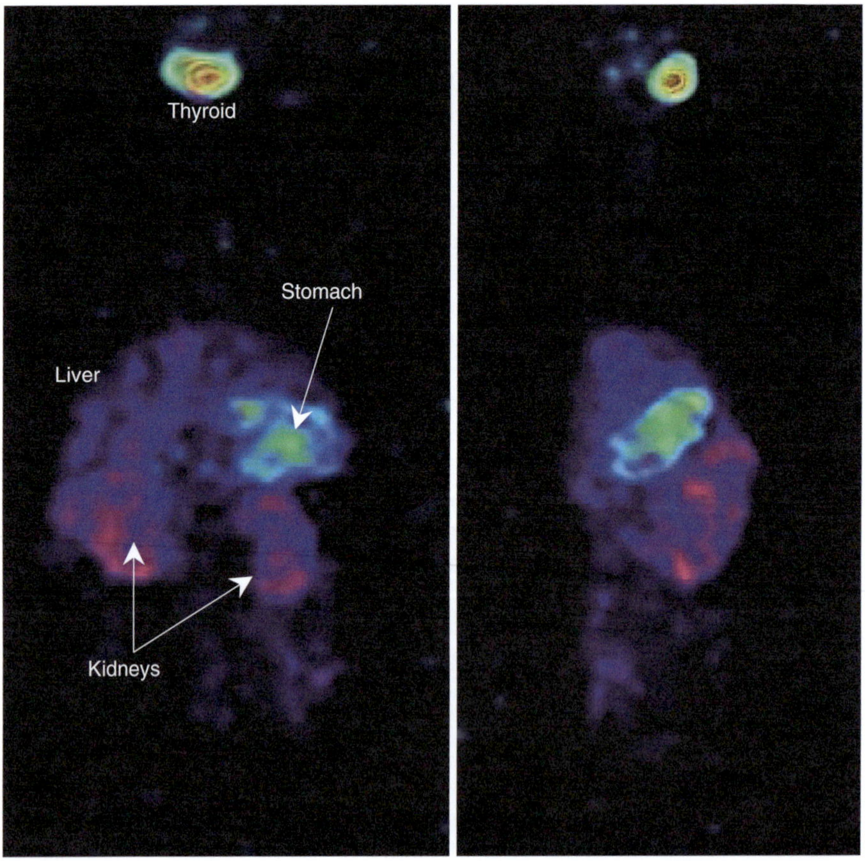


**Fig. 9.3** Dual-energy SPECT imaging of $^{125}$I-$^{aqa}$p5 and $^{123}$I-$^{AQA}$p5 peptides co-injected into a WT mouse. Coronal and sagittal views at 2 h postinjection show the distribution of $^{125}$I-$^{aqa}$p5 (pseudo-colored blue-red) is predominantly located in the liver and kidneys and the $^{123}$I-$^{AQA}$p5 radiotracer (pseudo-colored green-yellow) is confined to the stomach and thyroid

At 2 h pi, SPECT imaging revealed the presence of $^{aqa}$p5 in the liver and kidneys (blue-red color scale). In contrast, imaging of $^{AQA}$p5 revealed radioisotope within the stomach and thyroid (green-yellow color scale)—organs known to scavenge liberated radioiodide when peptide radiotracers undergo catabolism [33]. The two distinct distributions demonstrate the different catabolic mechanisms of these two, essentially identical, peptides in vivo and exemplify the power of dual-energy SPECT imaging to enable their simultaneous evaluation.

### 9.2.3 Evaluation of Radiolabeled $^{aqa}$p5 and $^{AQA}$p5 Peptides in AA Amyloid Mice

The transgenic murine model of AA amyloidosis that we use in this case study constitutively overexpresses the human interleukin-6 (hIL-6), which results in a chronic inflammatory process that results in the systemic deposition of abdominothoracic amyloid, most commonly seen in the liver, spleen, kidneys, heart, and intestines [25]. Since the structurally similar $^{aqa}$p5 and $^{AQA}$p5 are both amyloid-binding peptides, we anticipated that a direct comparison of the two, as radiotracers, in individual AA mice would result in similar distribution patterns. Dual-energy SPECT with contrast-enhanced CT imaging at 2 h pi confirmed the expected similar distribution in the amyloid-laden mouse with uptake of both radiotracers in the liver, spleen, and kidneys (Fig. 9.4). However, subtle differences were readily evident—notably, the lack of significant uptake of $^{123}$I-$^{AQA}$p5 in the intestine (Fig. 9.4c, d) as compared to $^{125}$I-$^{aqa}$p5 (Fig. 9.4a, b). These differences could only be documented in a dual-energy experiment.

Despite the ability of whole-body SPECT imaging to demonstrate the uptake of radiotracers in organs known to involve amyloid disease, these images do not demonstrate specific retention of the peptide in amyloid deposits. This is a critical step in the translational development of a novel radiotracer targeting a pathologic lesion in an experimental animal model. Therefore, to ensure that the radioactivity within these organs is associated with amyloid-specific retention of the radiotracer, we routinely perform microautoradiography on samples of each tissue [34]. This type of tissue analysis requires days of treatment making assessment of $^{123}$I-$^{AQA}$p5 impossible due to its 13-h half-life. However, a parallel imaging study can easily be performed using $^{125}$I-$^{AQA}$p5 and $^{125}$I-$^{aqa}$p5, to assess the former using this technique. Analysis of tissues from the $^{125}$I-$^{aqa}$p5-injected mouse revealed amyloid-specific binding of the radiotracer as evidenced by the co-localization of black silver grains (indicative of the presence of radioiodinated peptide in the autoradiograph [ARG]) with green birefringent amyloid in a Congo red (CR)-stained consecutive tissue section (Congo red is a definitive histological stain for amyloid, and green birefringence in stained tissues is pathognomonic for the pathology). Additionally, this technique readily indicated no off-target retention of $^{125}$I-$^{aqa}$p5 in the amyloid-free regions of the organs and provided evidence of the renal clearance of the peptide, as expected (Fig. 9.5, WT mouse kidney sample).

Dual-energy SPECT imaging using contrast-enhanced CT in the abdomen affords precise visualization of the radiotracers in abdominal organs and tissues;

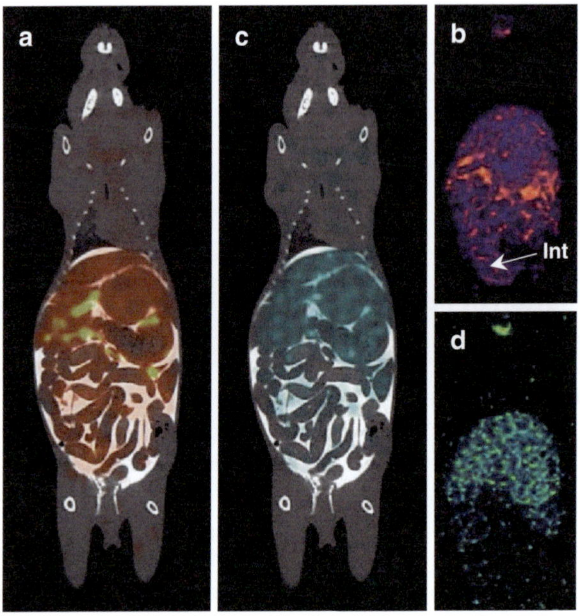


**Fig. 9.4** Dual-energy SPECT/CT imaging of $^{125}$I-$^{aqa}$p5 and $^{123}$I-$^{AQA}$p5 peptides in an individual mouse with AA amyloidosis. (**a**) 2D SPECT/CT (with contrast) of $^{125}$I-$^{aqa}$p5 (pseudo-colored red, sagittal plane). (**b**) 3D SPECT representation of $^{125}$I-$^{aqa}$p5. (**c**) 2D SPECT/CT (with contrast) of $^{123}$I-$^{AQA}$p5 (pseudo-colored blue). (**d**) 3D SPECT representation of $^{123}$I-$^{AQA}$p5 in the same mouse. When amyloid is present, the two radiotracers provide similar distribution; however, $^{125}$I-$^{aqa}$p5 also revealed uptake in the intestines (Int)

indeed, without the contrast-enhanced CT for anatomical reference (as shown in Fig. 9.4b, d), it would be challenging to discern organ boundaries, making analysis of the images extremely difficult (see Fig. 9.2). Intraperitoneal injection of an iodinated contrast agent, such as Iohexol™ (a 50% v:v solution), can be readily used for this purpose. This technique employs a relatively inexpensive contrast medium that can be injected in a volume of 500 μL and preserves the integrity of the tail vein that may be required for injection of radiotracers. With guidance from the contrast-enhanced CT, we were able to draw accurate regions of interest (ROIs) on organs which were then mapped onto the SPECT image data allowing precise measurements of tissue-associated radioactivity for both $^{125}$I-$^{aqa}$p5 and $^{123}$I-$^{AQA}$p5. This technique allows quantification of the two radiotracers within the same ROI in an individual mouse. These measurements (expressed in this case as mean voxel intensities) can then be compared with standard biodistribution measurements obtained by measuring radioactivity in tissues harvested at necropsy (often expressed as percent injected dose per gram of tissue)—obviously a terminal step in any imaging protocol. A comparison of the biodistribution measurements obtained from measuring radioactivity in tissue samples with those obtained directly from the analysis of

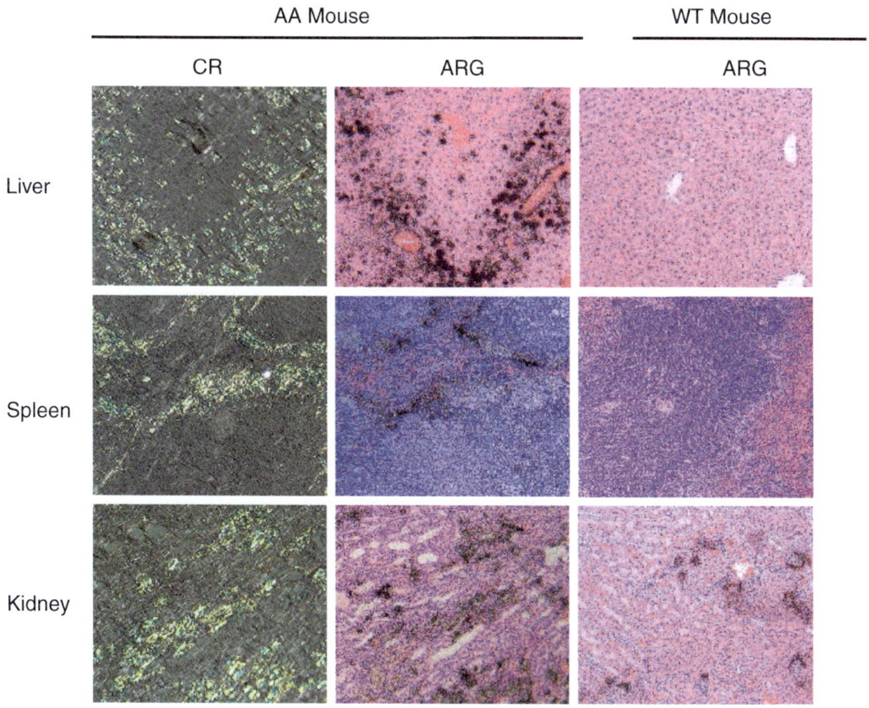


**Fig. 9.5** Congo red staining and microautoradiography of mouse tissues are used to visualize pathology and radiotracer within the tissues. Left panel shows green Congo red (CR) birefringence indicating amyloid deposition in the liver, spleen, and kidney. The center panel shows microautoradiography (ARG) of $^{125}$I-$^{aqa}$p5 (black dots) counterstained with hematoxylin and eosin for tissue visualization in a consecutive tissue slice. The peptide distribution in the ARG corresponds with amyloid seen in the CR images (left panel). The right panel shows $^{125}$I-$^{aqa}$p5 in WT tissues; no accumulation is seen in the liver or spleen, but the kidney reveals excretion of the peptide in Bowman's capsules and tubules

ROIs on the SPECT images revealed similar relative quantitation of each peptide radiotracer in the AA mice, due to their comparable amyloid-targeting efficacy (Figs. 9.6a, b). Similar analysis of the peptides in WT mice also corresponded with the image data in that $^{125}$I-$^{aqa}$p5 was seen predominantly in the kidneys, and free radioiodide liberated during catabolism of $^{123}$I-$^{AQA}$p5 was detected in the stomach (Fig. 9.6c, d). The thyroid, which also scavenges free iodide, is too small to dissect from mice to obtain accurate biodistribution data. Given these results, dual-energy SPECT/CT imaging can be considered a powerful tool for the quantitative comparison of two (or potentially more) independent radiotracers in individual mice, allowing detailed comparative analysis studies to be performed using small numbers of animals and accounting for the inherent biology variability of complex disease systems in animals.

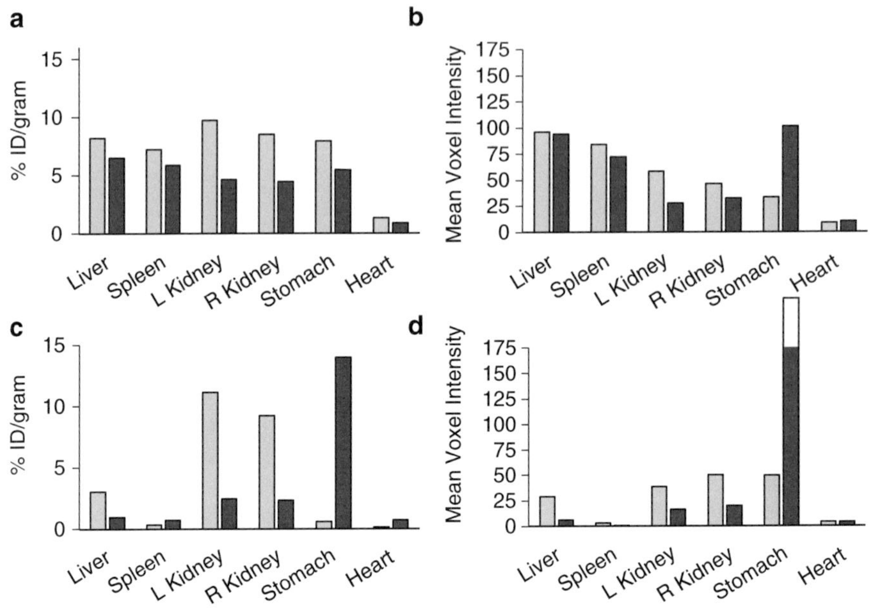


**Fig. 9.6** Tissue biodistribution measurements and image analyses of both $^{125}$I-$^{aqa}$p5 and $^{123}$I-$^{AQA}$p5 peptides in a single representative AA and WT mouse. Tissue biodistribution measurements of $^{125}$I-$^{aqa}$p5 (light gray) and $^{123}$I-$^{AQA}$p5 (dark gray) in an AA mouse (**a**) and WT mouse (**c**). Radiotracer distribution was similar in AA mice and varied in WT mice. Image analyses of both radiotracers in an AA (**b**) and WT (**d**) mouse

## 9.3 Materials and Methods

### 9.3.1 Radiotracer Preparation

Peptides $^{aqa}$p5 and $^{AQA}$p5 were synthesized commercially (AnaSpec, Fremont, CA), purified by reverse-phase HPLC and analyzed by mass spectrometry to ensure their integrity [18]. Peptide $^{aqa}$p5 was radiolabeled with $^{125}$I using 20 μg chloramine T followed by 20 μg sodium metabisulfite to quench the reaction. Radioiodination of peptide $^{AQA}$p5 with $^{123}$I was achieved using 1 mg/mL Iodogen (Pierce Chemical, Rockford, IL) oxidation followed with 2 mg/mL ascorbic acid reduction [33]. In both cases, radioiodine incorporated in the lone tyrosine amino acid in the sequence. Both radiolabeled peptides were purified from free radioiodide by gel filtration using a Sephadex G-25 size exclusion matrix (PD10, GE Healthcare). Fractions were collected under gravity, and gamma-counting was used to identify peptide in fractions with the highest radioactivity, which were then pooled. Radiochemical purity was assessed with SDS-PAGE analyzed by phosphor imaging (Cyclone Storage Phosphor System, PerkinElmer, Shelton, CT).

### 9.3.2 SPECT/CT Data Acquisition

Dual-energy SPECT/CT experiments were performed in individual AA or WT mice using an Inveon trimodality SPECT/PET/CT platform [35] (Siemens Preclinical Solutions, Knoxville, TN) with Inveon Acquisition Workplace (v.2) image reconstruction software. Although image acquisition parameters presented here are specific for this imaging platform, all preclinical SPECT multimodality imaging systems are capable of capturing and analyzing dual-energy SPECT/CT data. Detailed methods have been previously published [18]; however, briefly, AA ($n = 2$) and WT ($n = 2$) mice were each administered both radiotracers (113 μCi $^{125}$I-$^{aqa}$p5 and 46 μCi $^{123}$I-$^{AQA}$p5 mixed in a 200 μL volume) simultaneously in the lateral tail vein, and at 2 h pi, mice received ~0.5 mL of 1:1 dilution of Iohexol™ in PBS i.p. 1 min before being euthanized by an isoflurane inhalation overdose. An approximate 2:1 ratio for $^{125}$I- and $^{123}$I-labeled radiotracer, respectively, reduces spillover effects from high-energy photons emitted by $^{123}$I. SPECT data for the low- and high-energy isotopes were obtained sequentially with 60, 16-second projections in 1.5 revolutions using a 1-mm-diameter aperture and a five-pinhole "whole mouse body" collimator positioned 30 mm from the center of the field of rotation. Data were reconstructed using a point spread function model with a 3D maximum a priori algorithm (16 iterations, 6 subsets, $\beta = 1$) onto an 88 × 88 × 312 matrix with 0.5 mm isotropic voxels with appropriate scatter and attenuation correction applied.

CT data were acquired after the SPECT images using an 80 kVp X-ray voltage with a 500 μA anode current. Two bed positions were used with 240 ms exposure and 361 projections over a 360° rotation with binning at 4. Reconstruction was achieved with a Feldkamp-filtered cone beam algorithm onto a 256 × 256 × 603 matrix with 211.4 μm voxels.

### 9.3.3 Necropsy and Biodistribution

Organs (liver, spleen, kidneys, stomach, and heart) were harvested from each mouse at necropsy, and a small volume of tissue was placed into tared vials and weighed. Radioactivity was measured using an automated Wizard 3 gamma counter (1480 Wallac Gamma Counter, PerkinElmer) using low- and high-energy windows for $^{125}$I and $^{123}$I detection, respectively. Crossover radioactivity detected by the gamma counter (i.e., low-energy photons from $^{123}$I that appear in the $^{125}$I energy window) was accounted for by manually applying a 41.5% crossover correction, and the $^{125}$I values were reduced accordingly. Biodistribution of the peptides from this technique was expressed as percent injected dose per gram of tissue (%ID/g).

### 9.3.4 Congo Red Staining and Microautoradiography

Histological evaluation of tissue samples by Congo red (CR) and microautoradiography was performed using 6-μm-thick tissue sections cut from formalin-fixed, paraffin-embedded tissues which were harvested from mice at necropsy. For CR, tissues were placed on slides and stained with alkaline CR solution for 1 h at RT followed by Mayer's hematoxylin counterstain for 2 min. For microautoradiography, slides were dipped in Kodak NTB-2 emulsion, stored in the dark and developed after a 96 h exposure before being counterstained with hematoxylin. Tissues were examined using a Leica DM500 light microscope (Leica, Buffalo Grove, IL) with cross-polarizing filters (for CR), and digital images were obtained using a cooled CCD camera (SPOT RT-Slider; Diagnostic Instruments, Sterling Heights, MI) [18, 31].

### 9.3.5 Image Analysis

SPECT and contrast-enhanced CT image data were co-registered using the Inveon Research Workplace visualization software (Siemens Preclinical). From the CT image, small, spherical three-dimensional regions of interest were drawn over the liver, spleen, stomach, kidneys, and heart of amyloid-laden and WT mice, taking care to avoid regions thought to contain large blood vessels (in the liver). These volumes were then mapped to each SPECT dataset, and the mean voxel intensity of each radiotracer within the ROI was recorded. A spillover correction factor (5.9%) was applied to the $^{125}$I image data to account for the presence of additional low-energy photons arising from the $^{123}$I isotope. This correction factor for dual-energy image analyses was previously determined [24]. Additionally, the image data were appropriately scaled for differences in the injected dose.

## 9.4 Summary

Dual-energy, or indeed multi-energy, SPECT/CT imaging is a valuable technique that can be used to directly compare two or more suitably labeled radiotracers in vivo. Using both SPECT and contrast-enhanced CT modalities allows for precise anatomical guidance and enhanced accuracy of image analysis. The strength of this technique lies in the ability to perform comparative effectiveness studies of radiotracers in individual animals, thereby minimizing subject-to-subject variability which is an inherent complication in complex biological systems, particularly when disease models are used. In our exemplary study, we were able to quantitatively compare two radioiodinated peptides which behave differently in WT mice but have similar, although distinct, distribution patterns in mice with systemic AA amyloidosis. In the preclinical setting, dual-energy SPECT/CT provides a powerful tool in the development of imaging agents and other reagents designed to target pathological or physiological events, allowing for the direct comparison of reagents in vivo and thereby providing a pathway to identifying the optimal reagents for translation to clinical studies.

## References


1. Levin CS. Primer on molecular imaging technology. Eur J Nucl Med Mol Imaging. 2005;32(Suppl 2):S325–45. https://doi.org/10.1007/s00259-005-1973-y.
2. Ziegler S. PET and SPECT. In: Kiessling F, Pichler B, editors. Small animal imaging. Berlin: Springer; 2011. p. 231–6.
3. Madsen MT. Recent advances in SPECT imaging. J Nucl Med. 2007;48(4):661–73.
4. Bazanez-Borgert M. Basics of SPECT, PET and PET/CT imaging: JASS; 2006.
5. Lee CH, Goo JM, Ye HJ, Ye SJ, Park CM, Chun EJ, Im JG. Radiation dose modulation techniques in the multidetector CT era: from basics to practice. Radiographics. 2008;28(5):1451–9. https://doi.org/10.1148/rg.285075075.
6. Keat N. Report 05016 CT scanner automatic exposure control systems. 2005. http://www.impactscan.org/reports/Report05016. Accessed 2 Feb 2018.
7. Levine CD, Aizenstein O, Lehavi O, Blachar A. Why we miss the diagnosis of appendicitis on abdominal CT: evaluation of imaging features of appendicitis incorrectly diagnosed on CT. AJR Am J Roentgenol. 2005;184(3):855–9. https://doi.org/10.2214/ajr.184.3.01840855.
8. Aide N, Kinross K, Beauregard JM, Neels O, Potdevin T, Roselt P, Dorow D, Cullinane C, Hicks RJ. A dual radiologic contrast agent protocol for $^{18}$F-FDG and $^{18}$F-FLT PET/CT imaging of mice bearing abdominal tumors. Mol Imaging Biol. 2011;13(3):518–25. https://doi.org/10.1007/s11307-010-0378-x.
9. Thaiss WM, Sauter AW, Bongers M, Horger M, Nikolaou K. Clinical applications for dual energy CT versus dynamic contrast enhanced CT in oncology. Eur J Radiol. 2015;84(12):2368–79. https://doi.org/10.1016/j.ejrad.2015.06.001.
10. Forghani R, De Man B, Gupta R. Dual-energy computed tomography: physical principles, approaches to scanning, usage, and implementation: Part 2. Neuroimaging Clin N Am. 2017;27(3):385–400. https://doi.org/10.1016/j.nic.2017.03.003.
11. Silva AC, Morse BG, Hara AK, Paden RG, Hongo N, Pavlicek W. Dual-energy (spectral) CT: applications in abdominal imaging. Radiographics. 2011;31(4):1031–46.; ; discussion 1047–50. https://doi.org/10.1148/rg.314105159.
12. Histed SN, Lindenberg ML, Mena E, Turkbey B, Choyke PL, Kurdziel KA. Review of functional/anatomical imaging in oncology. Nucl Med Commun. 2012;33(4):349–61. https://doi.org/10.1097/MNM.0b013e32834ec8a5.
13. Mariani G, Bruselli L, Kuwert T, Kim EE, Flotats A, Israel O, Dondi M, Watanabe N. A review on the clinical uses of SPECT/CT. Eur J Nucl Med Mol Imaging. 2010;37(10):1959–85. https://doi.org/10.1007/s00259-010-1390-8.
14. Bernsen MR, Vaissier PE, Van Holen R, Booij J, Beekman FJ, de Jong M. The role of preclinical SPECT in oncological and neurological research in combination with either CT or MRI. Eur J Nucl Med Mol Imaging. 2014;41(Suppl 1):S36–49.
15. Buck AK, Nekolla S, Ziegler S, Beer A, Krause BJ, Herrmann K, Scheidhauer K, Wester HJ, Rummeny EJ, Schwaiger M, Drzezga A. Spect/Ct. J Nucl Med. 2008;49(8):1305–19. https://doi.org/10.2967/jnumed.107.050195.
16. Seo Y, Mari C, Hasegawa BH. Technological development and advances in single-photon emission computed tomography/computed tomography. Semin Nucl Med. 2008;38(3):177–98. https://doi.org/10.1053/j.semnuclmed.2008.01.001.
17. Ma KH, Huang WS, Chen CH, Lin SZ, Wey SP, Ting G, Wang SD, Liu HW, Liu JC. Dual SPECT of dopamine system using [$^{99m}$Tc]TRODAT-1 and [$^{123}$I]IBZM in normal and 6-OHDA-lesioned formosan rock monkeys. Nucl Med Biol. 2002;29(5):561–7.
18. Martin EB, Williams A, Richey T, Stuckey A, Heidel RE, Kennel SJ, Wall JS. Comparative evaluation of p5+14 with SAP and peptide p5 by dual-energy SPECT imaging of mice with AA amyloidosis. Sci Rep. 2016;6:22695. https://doi.org/10.1038/srep22695.
19. Meikle SR, Kench P, Kassiou M, Banati RB. Small animal SPECT and its place in the matrix of molecular imaging technologies. Phys Med Biol. 2005;50(22):R45–61. https://doi.org/10.1088/0031-9155/50/22/R01.

20. Nakazawa A, Ikeda K, Ito Y, Iwase M, Sato K, Ueda R, Dohi Y. Usefulness of dual $^{67}$Ga and $^{99m}$Tc-sestamibi single-photon-emission CT scanning in the diagnosis of cardiac sarcoidosis. Chest. 2004;126(4):1372–6. https://doi.org/10.1378/chest.126.4.1372.
21. Sanchez-Crespo A, Petersson J, Nyren S, Mure M, Glenny RW, Thorell JO, Jacobsson H, Lindahl SG, Larsson SA. A novel quantitative dual-isotope method for simultaneous ventilation and perfusion lung SPET. Eur J Nucl Med Mol Imaging. 2002;29(7):863–75. https://doi.org/10.1007/s00259-002-0803-8.
22. Wall JS, Richey T, Williams A, Stuckey A, Osborne D, Martin E, Kennel SJ. Comparative analysis of peptide p5 and serum amyloid P component for imaging AA amyloid in mice using dual-isotope SPECT. Mol Imaging Biol. 2012;14(4):402–7. https://doi.org/10.1007/s11307-011-0524-0.
23. Weinmann P, Faraggi M, Moretti JL, Hannequin P. Clinical validation of simultaneous dual-isotope myocardial scintigraphy. Eur J Nucl Med Mol Imaging. 2003;30(1):25–31. https://doi.org/10.1007/s00259-002-0995-y.
24. Lee S, Gregor J, Kennel SJ, Osborne DR, Wall J. GATE validation of standard dual energy corrections in small animal SPECT-CT. PLoS One. 2015;10(4):e0122780. https://doi.org/10.1371/journal.pone.0122780.
25. Wechalekar AD, Gillmore JD, Hawkins PN. Systemic amyloidosis. Lancet. 2015; https://doi.org/10.1016/S0140-6736(15)01274-X.
26. Hazenberg BP, van Rijswijk MH, Piers DA, Lub-de Hooge MN, Vellenga E, Haagsma EB, Hawkins PN, Jager PL. Diagnostic performance of $^{123}$I-labeled serum amyloid P component scintigraphy in patients with amyloidosis. Am J Med. 2006;119(4):355 e315–24. https://doi.org/10.1016/j.amjmed.2005.08.043.
27. Dorbala S, Vangala D, Semer J, Strader C, Bruyere JR Jr, Di Carli MF, Moore SC, Falk RH. Imaging cardiac amyloidosis: a pilot study using (18)F-florbetapir positron emission tomography. Eur J Nucl Med Mol Imaging. 2014;41(9):1652–62. https://doi.org/10.1007/s00259-014-2787-6.
28. Law WP, Wang WY, Moore PT, Mollee PN, Ng AC. Cardiac amyloid imaging with $^{18}$F-Florbetaben PET: a pilot study. J Nucl Med. 2016;57(11):1733–9. https://doi.org/10.2967/jnumed.115.169870.
29. Osborne DR, Acuff SN, Stuckey A, Wall JS. A routine PET/CT protocol with streamlined calculations for assessing cardiac amyloidosis using (18)F-Florbetapir. Front Cardiovasc Med. 2015;2:23. https://doi.org/10.3389/fcvm.2015.00023.
30. Rossi P, Tessonnier L, Frances Y, Mundler O, Granel B. $^{99m}$Tc DPD is the preferential bone tracer for diagnosis of cardiac transthyretin amyloidosis. Clin Nucl Med. 2012;37(8):e209–10. https://doi.org/10.1097/RLU.0b013e318248512c.
31. Wall JS, Martin EB, Richey T, Stuckey AC, Macy S, Wooliver C, Williams A, Foster JS, McWilliams-Koeppen P, Uberbacher E, Cheng X, Kennel SJ. Preclinical validation of the heparin-reactive peptide p5+14 as a molecular imaging agent for visceral amyloidosis. Molecules. 2015;20(5):7657–82. https://doi.org/10.3390/molecules20057657.
32. Martin EB, Williams A, Richey T, Wooliver C, Stuckey A, Foster JS, Kennel SJ, Wall JS. Evaluation of the effect of D-amino acid incorporation into amyloid-reactive peptides. J Transl Med. 2017;15(1):247. https://doi.org/10.1186/s12967-017-1351-0.
33. Martin EB, Kennel SJ, Richey T, Wooliver C, Osborne D, Williams A, Stuckey A, Wall JS. Dynamic PET and SPECT imaging with radioiodinated, amyloid-reactive peptide p5 in mice: a positive role for peptide dehalogenation. Peptides. 2014;60:63–70. https://doi.org/10.1016/j.peptides.2014.07.024.
34. Wall JS, Paulus MJ, Gleason S, Gregor J, Solomon A, Kennel SJ. Micro-imaging of amyloid in mice. Methods Enzymol. 2006;412:161–82. https://doi.org/10.1016/S0076-6879(06)12011-X.
35. Magota K, Kubo N, Kuge Y, Nishijima K, Zhao S, Tamaki N. Performance characterization of the Inveon preclinical small-animal PET/SPECT/CT system for multimodality imaging. Eur J Nucl Med Mol Imaging. 2011;38(4):742–52. https://doi.org/10.1007/s00259-010-1683-y.

# 10 Multimodality Imaging in Small Animal Radiotherapy

Christian Vanhove and Stefaan Vandenberghe

## 10.1 Introduction

Cancer is a devastating disease with a rapidly increasing incidence, mainly due to the ageing population. In 2017, it was estimated that the predicted number of cancer deaths in the European Union was 1,373,500, compared to 1,333,400 in 2012 (+3%) [1]. Despite improvements in prevention and treatment, cancer remains one of the most important causes of morbidity and mortality, leading to high economic costs. Together with surgery and systemic treatment, radiotherapy is a key part of cancer treatment, both in the curative and in the palliative setting. About half of cancer patients will receive radiotherapy as part of their anticancer treatment. Multimodality imaging is the cornerstone of radiation treatment planning and allows to deliver a high radiation dose to the tumour, while at the same time keeping the dose in the surrounding tissue sufficiently low to prevent side effects. In Fig. 10.1, a typical workflow is shown when radiotherapy is involved during treatment. Multimodality imaging is used at multiple time points: at the stage of disease diagnosis, for the planning of the treatment, repeatedly during treatment, and after treatment to monitor anticancer treatment response. In case of a multi-fraction treatment, imaging during the course of treatment enables corrective action in a feedback loop, known as adaptive radiotherapy. This also allows alteration of the treatment in tumours that are not responding to radiotherapy.

The past 40-odd years have seen a great progress in imaging technologies. These images enable to better localize the desired targets for radiation. It was the pioneering work of Sir Godfrey Hounsfield that resulted in the introduction of computed tomography (CT) and allowed for the first time to non-invasively visualize a tumour

C. Vanhove (✉) · S. Vandenberghe
Institute Biomedical Technology, Medical Imaging and Signal Processing, University of Ghent, Ghent, Belgium
e-mail: Christian.Vanhove@UGent.be





C. Kuntner-Hannes, Y. Haemisch (eds.), *Image Fusion in Preclinical Applications*, https://doi.org/10.1007/978-3-030-02973-9_10

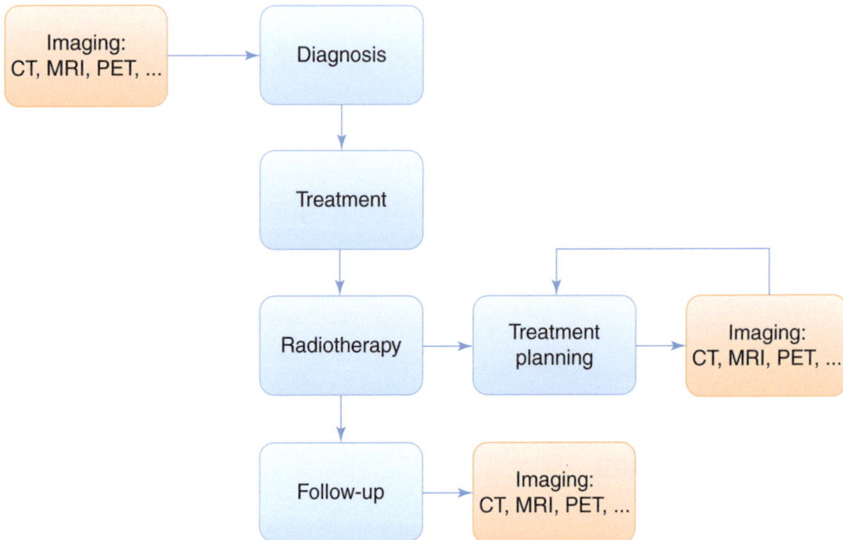


**Fig. 10.1** Radiotherapy workflow indicating the role of imaging at multiple stages. A feedback loop based on imaging information allows adaptation of the therapy to optimize the outcome

in the three spatial dimensions [2], opening the era of three-dimensional conformal radiotherapy. In three-dimensional conformal radiotherapy, patient-specific target selection can be performed based on image information, together with the identification of sensitive tissues. Based on this one can develop treatment plans that conform to the target and avoid sensitive tissues as best as possible. At first, the creation of treatment plans was a process of trial and error and required the optimization of a number of parameters, such as the number of radiation beams, the beam size, the beam weights, and the beam arrangement. Based on CT information, these parameters were used to accurately model the behaviour of the radiation beams as they travel through the body [3–5]. Recently, three-dimensional conformal radiotherapy has been replaced by intensity-modulated radiotherapy, where dynamic multi-leaf collimators in combination with variable intensities of the radiation beam are used to allow even greater control over the shape of the dose distribution [3]. Intensity-modulated radiotherapy allows almost unlimited degrees of freedom to shape the radiation beams. As a result, inverse planning is required to (automatically) calculate the desired treatment plan. This process requires defining the target dose for tumour and sensitive tissues, and a mathematical optimization algorithm is then used to create a treatment plan that best matches the input criteria. This allows the delivery of very complex treatment plans, only limited by the physics that governs photons. As radiation treatment is a delicate balance between tumour control and normal tissue toxicity, further treatment plan optimization can only be reached when we obtain greater knowledge of the target and a better understanding of normal tissue complications. However, as mentioned by Hoffmann et al. [6], in current treatment plans, the compromise between the dose delivered to tumour and normal tissue is 'frozen', based on what is considered as the best trade-off for a specific patient population. Moreover, in the current clinical radiotherapy practice, the tumours are generally irradiated with a spatially uniform dose. It is known, however, that some tumour regions need a higher dose to be destroyed than other regions [7, 8] and the hypothesis is that a non-uniform dose distribution should be used to improve treatment outcome. The information of intra-tumour inhomogeneity comes from imaging, and it is hypothesized that advances in multimodality image-guided radiotherapy will lead to improved cancer cure rates. Imaging modalities, such as magnetic resonance imaging (MRI) and positron emission tomography (PET), can visualize intra-tumoural biological heterogeneity [9] and have the potential to improve current treatment strategies.

## 10.2 Small Animal Radiation Research

For many decades, animal radiation studies were mostly performed using fairly crude experimental setups with radiation fields that did not conform only to the desired target [10]. Commonly, these experiments were done on devices intended for human patient use, and the radiation sources employed were often producing megavolt (MV) X-rays that have several characteristics that are unsuitable for irradiating small targets in small animals [11]. A MV photon beam exhibits dose build-up at the air-tissue interface in the entrance region of the beam. The extent of this build-up region corresponds roughly to the order of the animal size itself. This makes it very challenging to deliver a uniform dose to a tumour. Another issue is the beam penumbra, which for MV photon beams may extend several millimetres beyond the target, leading to unacceptable dose distributions in small structures.

However, to implement changes to the existing clinical standard of care, research must be conducted to develop alternative treatment strategies. Therefore, a novel approach in radiotherapy is the introduction of preclinical precision image-guided radiation research. Tumour-bearing small animal models (mostly mice and rats) are used to investigate the efficacy of complex radiation patterns, possibly combined with other treatment agents (e.g. angiogenesis inhibitors or radiosensitizers) that would otherwise be unethical to investigate on patients. Small animal radiation research can also play an important role in assessing radiation response of normal tissue and in investigating radioprotective agents. Moreover, preclinical radiation research allows for studies under controlled experimental conditions using large cohorts and delivering accelerated results due to the shorter lifespans of rodents. Consequently, there is an emerging consensus that novel combinations of imaging and therapy regimen should first be investigated in a preclinical research environment offering precision irradiation and multimodality imaging. The preclinical findings should then be translatable into a clinical trial in a much faster and more efficient way than in current practice.

As a result, precision image-guided small animal radiation research platforms were developed [11, 12]. These platforms make use of kilo-voltage (kV) X-ray sources to avoid dose build-up and to obtain extremely sharp penumbras. They typically integrate:

- A kV X-ray source that is used for imaging and radiation treatment.
- A computer-controlled stage for animal positioning.
- A rotational gantry assembly to allow radiation delivery from various angles.
- A collimating system to shape the radiation beam.

This is similar to modern human radiotherapy practice and enables a wide variety of preclinical experiments, such as the synergy of radiation with other therapies, complex radiation schemes, and image-guided sub-target boost studies. In Fig. 10.2, an example of a small animal radiation research platform is shown (SARRP, XStrahl, Surrey, UK).

Treatment planning on these small animal radiation research platforms is based on CT, which is equivalent to human planning systems and currently the gold standard for radiation planning [4, 5]. For CT imaging an on-board X-ray detector is used in combination with the same kV X-ray tube that is used during treatment (Fig. 10.2). CT imaging is preferred as it allows for accurate beam positioning and provides electron density information necessary for individual radiation dose calculations. However, the CT systems installed on these research platforms are based on

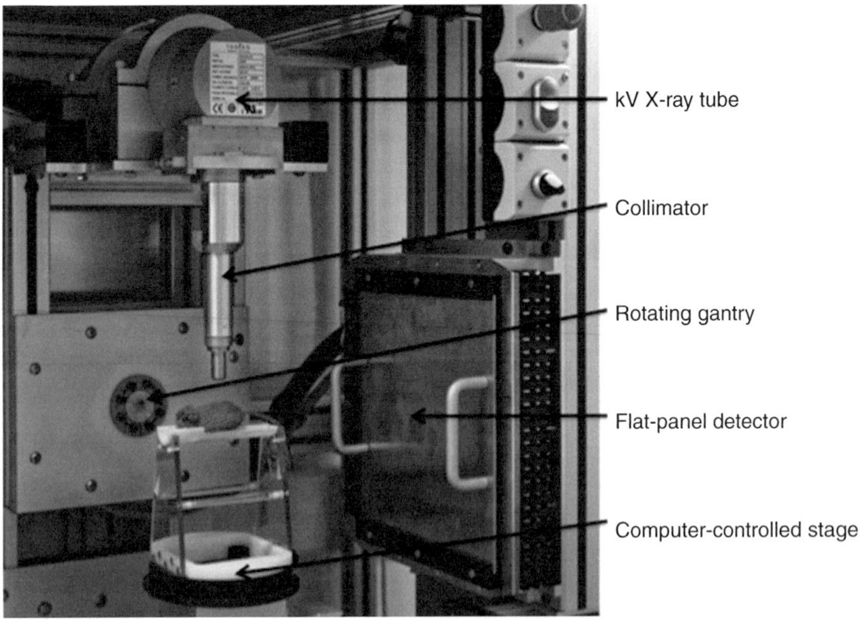


**Fig. 10.2** Small animal radiation research platform integrating a kV X-ray tube, a rotating gantry, a computer-controlled stage, a collimating system to shape the beam, and a flat-panel CT detector

the cone-beam (CB) geometry, instead of the spiral (slice-based) CT geometry used in human systems. CB-CT is hampered by low soft tissue contrast and high background noise as a result of the large amount of scatter due to a high scatter-to-primary ratio when no anti-scatter grid is used on these systems [13]. Although many investigators have shown that CB-CT can be extremely useful for guiding focal irradiation [14–16], it remains challenging to localize soft tissue targets on CB-CT images. To localize soft tissue targets more efficiently, CB-CT can be combined with other imaging modalities, where the alternative imaging technology is often used for target selection, and CB-CT is used for dose calculations and accurate beam positioning.

## 10.3 Multimodality Imaging in Small Animal Radiation Research

Currently, a large number of preclinical non-invasive in vivo imaging techniques are available that have the potential to provide more detailed information compared to CB-CT, such as optical imaging, MRI, and nuclear imaging techniques [17, 18]. Each of these modalities has intrinsic advantages and limitations when used in combination with small animal precision image-guided radiation research platforms. In the next three sections, these advantages and limitations will be discussed into more detail.

### 10.3.1 CB-CT Combined with Bioluminescence Imaging

Bioluminescence imaging (BLI) enables to directly visualize cancer cells when they are transfected with luciferase [19]. Related to image guidance, the major advantage of BLI is that it provides excellent signal-to-background ratios due to the negligible background signal. Moreover, it is a relatively inexpensive imaging technique, the short acquisition times lead to high throughput, it allows non-invasive monitoring of tumour progression, and the compact footprint enables to integrate BLI into a micro-irradiator. However, BLI suffers from absorption and scattering of visible light by tissue, which results in a nonlinear relationship between the true signal strength and the measured optical signal at the animal's surface, limiting the accuracy to localize a target in 3D. BLI is mostly available in planar mode, though bioluminescence tomography (BLT) is feasible, and is hampered by a limited spatial resolution [17].

Tuli et al. [20] demonstrated the feasibility of BLI-guided irradiation in an orthotopic mouse model of pancreatic cancer using an offline optical imager. Targeting accuracy was measured using a glass bulb, with 5 mm diameter and filled with bioluminescent cells, which were orthotopically implanted into the tail of the pancreas. The centroid of this bulb measured on the CB-CT of the micro-irradiator was used as a reference. Using planar optical images, a targeting accuracy of 3.5 mm could be achieved, which indicates the deviation from the centroid of the implanted bulb to a vertical line going through the maximum optical signal detected on the animal's surface. Accordingly, using only planar optical images, no accurate information of the depth position of the target can be provided. Therefore, the same group investigated the added value of offline BLT to guide irradiation [21]. Although an overall accuracy of approximately 1 mm could be achieved with BLT, the authors indicated that this was a best-case scenario because ex vivo mice were used during the experiment, minimizing the repositioning error when moving the animals from the optical imager to the CB-CT. They concluded that an integrated optical/CB-CT instrument is necessary to further reduce the targeting error.

Weersink et al. [22] were the first to integrate BLI into a micro-irradiator and concluded that a targeting accuracy of 1 mm can be achieved in rigid homogeneous phantoms, using planar optical images. The uncertainty of the depth position of the target was reduced by using a pair of parallel-opposed radiation beams.

Recently, a group from Johns Hopkins University [23, 24] introduced an integrated BLT/CB-CT system (Fig. 10.3) in a small animal radiation research platform. Using a novel reconstruction algorithm based on multispectral BLT [25, 26] and incomplete variables truncated conjugate gradients [27], an overall targeting accuracy of 1 mm could be obtained using phantoms and ex vivo mice experiments.

Improvements of these integrated multimodality systems should be further investigated. For example, the systems described by Zhang et al. [24] (Fig. 10.3) require manual docking of the BLT system, and the authors mention that efforts are in progress for an automatic system that will result in better mechanical reproducibility.

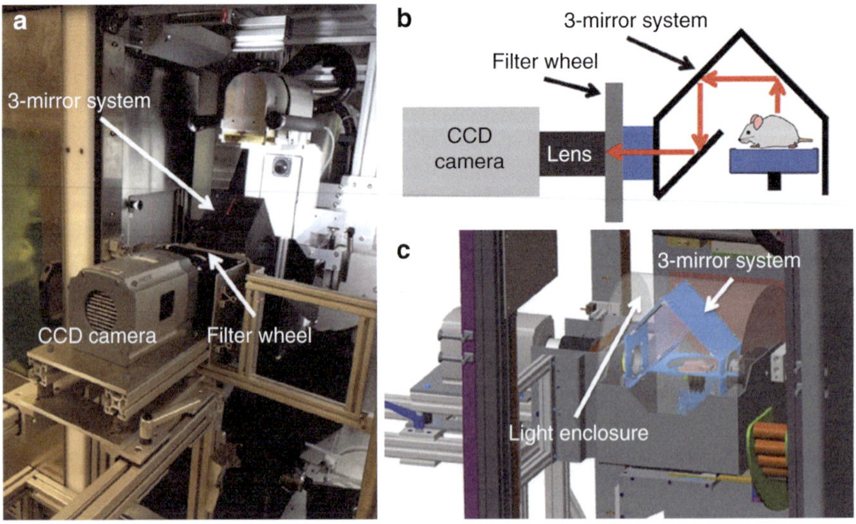


**Fig. 10.3** Bioluminescence tomography integrated into a small animal radiation research platform. Reprinted from [24] with permission from Elsevier

### 10.3.2 CB-CT Combined with MRI

Compared to CB-CT and CT, MRI provides vastly superior soft tissue contrast. This makes it much easier to visualize lesion boundaries that will result in a much better delineation of the target volume, helping to better irradiate the lesion and avoid surrounding tissue. An additional advantage is that MRI uses nonionizing radio-waves, unlike CT that is using ionizing radiation. The major disadvantages of MRI are the relatively long acquisition times, the significant investment in an MRI scanner and high operational costs. Moreover, integrating an MRI into a radiation platform is far from trivial, notwithstanding, clinical systems are currently under construction [28–30].

Because MRI scans alone cannot be used for dose planning, as they do not provide the required electron density information, combining MRI with CT data is increasingly used for radiotherapy planning in the clinic [31]. This combined CT/MRI dataset contains both the information required for targeting (MRI-based volumes) and for dose calculations (CT-based electron density). Obviously, correct registration between MRI and CT is required to obtain accurate treatment planning.

Preclinically, only a few studies have been published that are using MRI-based radiotherapy. Bolcaen et al. [32] successfully applied a combined CB-CT/MRI dataset for the irradiation of brain tumours in a F98 glioblastoma rat model [33] using a micro-irradiator. Rigid-body transformations in combination with a multimodality bed were used for image registration between MRI and CB-CT. Contrast-enhanced MR images were acquired to follow up tumour growth after orthotopic inoculation, to monitor treatment response, and to delineate the target volume during radiotherapy planning. Using three noncoplanar arcs, the prescribed dose could be delivered to 90% of the target volume, while minimizing the dose to normal brain tissue. The authors concluded that this combined CT/MRI-based workflow is a major step forward in bridging the gap between preclinical and clinical radiotherapy planning.

One of the more challenging aspects of CB-CT imaging relates to the radiation dose received by the animals, where measurements showed typical radiation doses in the range of 10–50 cGy for a single CB-CT [34]. The dose at which 50% of mice die within 30 days (LD50/30) is roughly 5 Gy, which means that a single CB-CT can represent as much as 10% of the LD50/30. This might become a very important issue when the therapeutic dose has to be delivered in multiple fractions spaced over time, where each individual irradiation requires a CB-CT for accurate animal positioning. Therefore, Gutierrez et al. [35] investigated the feasibility of a MRI-only-based workflow for radiotherapy planning of the rat brain that enables both accurate target delineation and accurate dose calculations using only MRI-based volumes. Moreover, the image registration process between CB-CT and MRI would become redundant using such an MRI-only-based workflow. Multiple MRI sequences were used to generate synthetic CT images that could be used for dose calculations (Fig. 10.4), because only one MRI volume was not sufficient to separate all major tissue types (air, soft tissue, bone) in the rat head. The synthetic CT images were

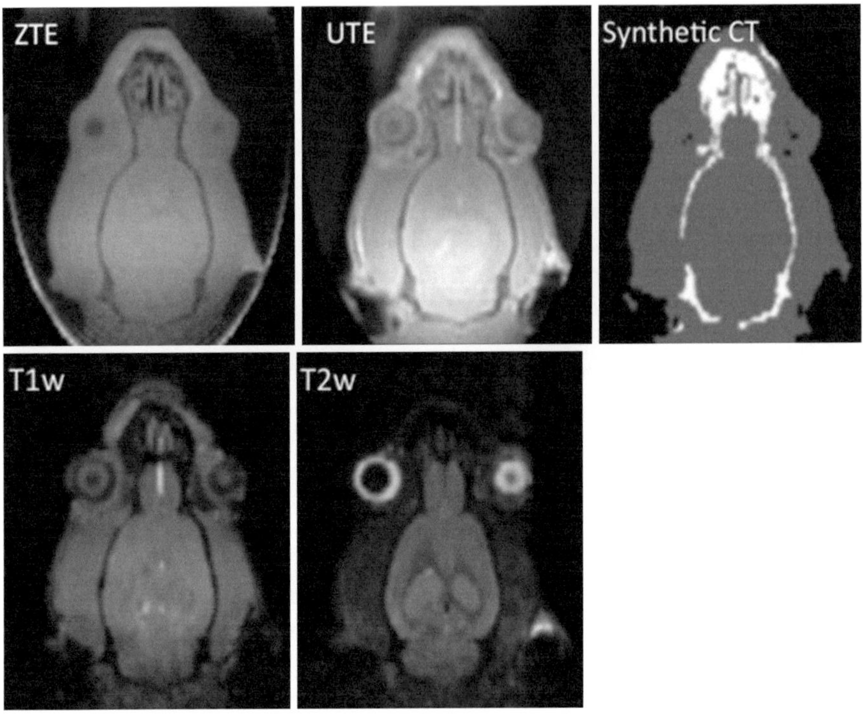


**Fig. 10.4** Coronal slices through the rat head using four different MRI sequences: zero echo time (ZTE), ultra-short echo time (UTE), T1-weighted (T1w), and T2-weighted (T2w). These images were used to generate a synthetic CT image for dose calculations

sufficiently similar to the segmented CB-CT images that are routinely used for radiotherapy planning on preclinical radiation research platforms. No significant differences were observed between CB-CT- and MRI-based dose calculations when more complex beam configurations (multiple beams) were used in the dose plan. The authors concluded that further research is required in the thoracic or abdominal region of small animals, where more tissue classes will be required to allow for accurate dose calculations compared to the rat head. Moreover, total MRI scan time might become an issue because of animal anaesthesia and throughput, and the proposed MRI-only-based workflow still requires the on-board CB-CT of the micro-irradiation for accurate animal positioning. For the latter, a solution should be found to ensure a common coordinate system between MR image space and micro-irradiator space, which is a non-trivial issue without the on-board CB-CT information. A possible solution is the use of digitally reconstructed radiographs (DRRs), extracted from the acquired MRI volumes, which may provide sufficient information for the purpose of image guidance.

### 10.3.3 CB-CT Combined with Nuclear Imaging

An advantage of nuclear imaging techniques, such as positron emission tomography (PET), is that these can be used to target metabolically highly active regions and have the potential to visualize intra-tumoural biological heterogeneity [9]. In 2000, Ling et al. [36] introduced the concept of biological target volume (BTV) by integrating anatomical and functional imaging into the radiotherapy workflow, leading to what they called multidimensional conformal radiotherapy. This offers the opportunity to improve dose targeting by delivering a non-uniform dose to a target region using, for example, PET images. The most widely used PET tracer for tumour staging and to monitor treatment response is fluor-18 ($^{18}$F)-labelled fluorodeoxyglucose (FDG), which visualizes the glucose metabolism [37]. However, PET can also be used to detect hypoxic tumour regions with tracers such as $^{18}$F-fluoromisonidazole (FMISO) [38]. Because tumour hypoxia is a known radioresistance factor, such information can be used to give an extra boost of radiation to hypoxic regions of the tumour. Despite that, the implementation of PET for radiotherapy planning is still under investigation in the clinic [39–41], and preclinical research might provide new insights for combining PET with radiotherapy. However, only a limited number of preclinical studies have been published. In 2007, Schütze et al. [42] investigated whether tumour heterogeneity in FDG uptake relates to radiation response. They showed that an increase of radiation dose had a greater effect on long-term recurrence-free survival in tumours with higher FDG uptake than in tumours with lower FDG uptake, supporting the hypothesis that pretreatment FDG-PET may provide useful information to deliver a non-uniform dose to the tumour. This hypothesis was evaluated by Trani et al. [43], who assessed the therapeutic efficacy of two different strategies to boost tumour sub-volumes with high FDG uptake: targeted dose escalation and dose redistribution. For targeted dose escalation, a 40% higher dose was delivered to a tumour region with high FDG uptake compared to the rest of the tumour. In the dose redistribution experiments, a 40% or 60% higher dose was delivered to a tumour region with high FDG, while a lower dose was delivered to the rest of the tumour. In contrast to what they expected, their results indicated that a decrease of radiation dose to tumour sub-volumes with low FDG uptake, while increasing the dose to high FDG uptake sub-volumes in a dose redistribution approach, might be detrimental for some dose levels. They concluded that their data are consistent with a hypothesis that tumour response depends on a minimum intra-tumoural dose. Thus further preclinical research is required before translating radiotherapy based on metabolic active regions into clinical trials (Fig. 10.5).

Evidently, PET-based radiotherapy planning requires correct registration with the planning CB-CT to obtain accurate treatment planning. This process can be simplified by using a multimodality bed to move the animal from the PET device to the micro-irradiator. In addition, the use of a hybrid PET-CT scanner can make the co-registration less difficult by using the CB-CT and the CT acquired in combination with the PET for image fusion.

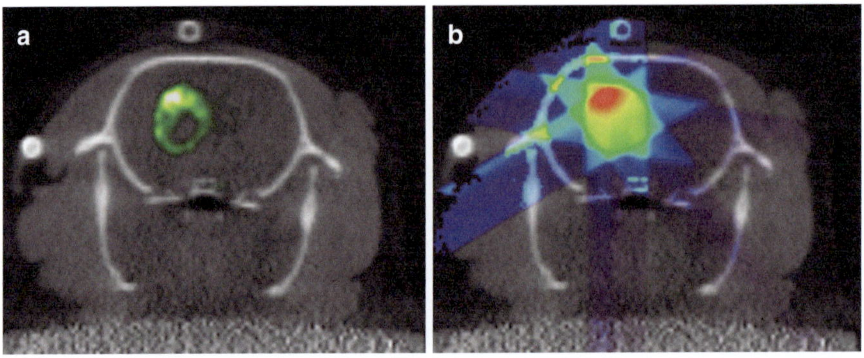


**Fig. 10.5** Example of PET-based treatment planning. (**a**) PET-CT image of the rat brain in a rat model of glioblastoma where a heterogeneous uptake of fluor-18-labelled fluorodeoxyglucose (FDG) can be observed in the tumour. (**b**) Treatment planning based on multiple beams to deliver a heterogeneous dose to the tumour. In this example, a larger dose is delivered to the part of the tumour with larger FDG uptake

Finally, compact preclinical PET scanners offering submillimetre spatial resolution are under development [44], and these devices might provide a very elegant solution to integrate PET into a small animal radiation platform.

## 10.4 Conclusion

To enable more accurate irradiation in small animal research, precision image-guided small animal radiation research platforms were developed. Similar to clinical systems, treatment planning on these platforms is based on CT. However, preclinical CT is hampered by very poor soft tissue contrast and high radiation doses. Consequently, multimodality imaging is the next logical step in the field of small animal precision image-guided radiotherapy.

## References

1. Malvezzi M, Carioli G, Bertuccio P, Boffetta P, Levi F, La Vecchia C, et al. European cancer mortality predictions for the year 2017, with focus on lung cancer. Ann Oncol [Internet]. 2017;28(5):1117–23. https://academic.oup.com/annonc/article-lookup/doi/10.1093/annonc/mdx033.
2. Petrik V, Apok V, Britton JA, Bell BA, Papadopoulos MC. Godfrey Hounsfield and the dawn of computed tomography. Neurosurgery. 2006;58(4):780–7.
3. Khan FM. The physics of radiation therapy. 4th ed. Philadelphia: Lippincott Williams and Wilkins; 2010. 531 p.
4. Aird EGA, Conway J. CT simulation for radiotherapy treatment planning. Br J Radiol. 2002;75(900):937–49.
5. Baker GR. Localization: conventional and CT simulation. Br J Radiol. 2006;79(Spec. No. 1):S36–49.

6. Hoffmann AL, Huizenga H, Kaanders JH. Employing the therapeutic operating characteristic (TOC) graph for individualised dose prescription. Radiat Oncol [Internet]. 2013;8(1):55. http://www.pubmedcentral.nih.gov/articlerender.fcgi?artid=3606307&tool=pmcentrez&rendertype=abstract.
7. Bentzen SM, Gregoire V. Molecular imaging-based dose painting: a novel paradigm for radiation therapy prescription. Semin Radiat Oncol [Internet]. 2011;21(2):101–10. https://doi.org/10.1016/j.semradonc.2010.10.001.
8. Diaz-Cano SJ. Tumor heterogeneity: mechanisms and bases for a reliable application of molecular marker design. Int J Mol Sci. 2012;13(2):1951–2011.
9. Metz S, Ganter C, Lorenzen S, Van Marwick S, Holzapfel K, Herrmann K, et al. Multiparametric MR and PET imaging of intratumoral biological heterogeneity in patients with metastatic lung cancer using voxel-by-voxel analysis. PLoS One. 2015;10(7):1–14.
10. Hamstra DA, Rice DJ, Anthony P, Oyedijo D, Ross BD, Rehemtulla A. Combined radiation and enzyme/prodrug treatment for head and neck cancer in an orthotopic animal model. Radiat Res. 1999;152(5):499–507.
11. Verhaegen F, Granton P, Tryggestad E. Small animal radiotherapy research platforms. Phys Med Biol. 2011;56(12):R55–83.
12. Butterworth KT, Prise KM, Verhaegen F. Small animal image-guided radiotherapy: status, considerations and potential for translational impact. Br J Radiol. 2015;88(1045):4–6.
13. Siewerdsen JH, Moseley DJ, Bakhtiar B, Richard S, Jaffray DA. The influence of antiscatter grids on soft-tissue detectability in cone-beam computed tomography with flat-panel detectors. Med Phys. 2004;31(12):3506–20.
14. Bolcaen J, Descamps B, Deblaere K, Boterberg T, De Vos F, Kalala Okito J-P, et al. F18-fluoromethylcholine (FCho), F18-fluoroethyltyrosine (FET) and F18-fluorodeoxyglucose (FDG) for the discrimination between high-grade glioma (HGG) and radiation necrosis (RN): a μPET study. Soc Nucl Med Annu Meet Abstr [Internet]. 2014;55(Suppl. 1):1379. http://jnumedmtg.snmjournals.org/cgi/content/meeting_abstract/55/1_MeetingAbstracts/1379.
15. Clarkson R, Lindsay PE, Ansell S, Wilson G, Jelveh S, Hill RP, et al. Characterization of image quality and image-guidance performance of a preclinical microirradiator. Med Phys [Internet]. 2011;38(2):845–56. http://www.pubmedcentral.nih.gov/articlerender.fcgi?artid=3188651&tool=pmcentrez&rendertype=abstract
16. Baumann BC, Benci JL, Santoiemma PP, Chandrasekaran S, Hollander AB, Kao GD, et al. An integrated method for reproducible and accurate image-guided stereotactic cranial irradiation of brain tumors using the small animal radiation research platform. Transl Oncol [Internet]. 2012;5(4):230–7. http://www.pubmedcentral.nih.gov/articlerender.fcgi?artid=3431032&tool=pmcentrez&rendertype=abstract.
17. James ML, Gambhir SS. A molecular imaging primer: modalities, imaging agents, and applications. Physiol Rev. 2012;92(2):897–965.
18. Cunha L, Horvath I, Ferreira S, Lemos J, Costa P, Vieira D, et al. Preclinical imaging: an essential ally in modern biosciences. Mol Diagn Ther. 2014;18(2):153–73.
19. Sato A, Klaunberg B, Tolwani R. In vivo bioluminescence imaging BLI: an overview. Comp Med. 2004;54(6):631–4.
20. Tuli R, Surmak A, Reyes J, Hacker-Prietz A, Armour M, Leubner A, et al. Development of a novel preclinical pancreatic cancer research model: bioluminescence image-guided focal irradiation and tumor monitoring of orthotopic xenografts. Transl Oncol [Internet]. 2012;5(2):77–84. https://doi.org/10.1593/tlo.11316%5Cnpapers2://publication/doi/10.1593/tlo.11316
21. Tuli R, Armour M, Surmak A, Reyes J, Iordachita I, Patterson M, et al. Accuracy of off-line bioluminescence imaging to localize targets in preclinical radiation research. Radiat Res. 2013;179(4):416–21.
22. Weersink RA, Ansell S, Wang A, Wilson G, Shah D, Lindsay PE, et al. Integration of optical imaging with a small animal irradiator. Med Phys [Internet]. 2014;41(10):102701. http://www.ncbi.nlm.nih.gov/pubmed/25281980.
23. Yang Y, Wang KK-H, Eslami S, Iordachita II, Patterson MS, Wong JW. Systematic calibration of an integrated X-ray and optical tomography system for preclinical radiation research.

Med Phys [Internet]. 2015;42:1710–20. http://scitation.aip.org/content/aapm/journal/medphys/42/4/10.1118/1.4914860.
24. Zhang B, Wang KK-H, Yu J, Eslami S, Tran PT, Iordachita I, et al. Bioluminescence tomography guided radiation therapy for preclinical research. Radiat Oncol Biol [Internet]. 2016;94(5):1–27. https://doi.org/10.1016/j.ijrobp.2015.11.039.
25. Dehghani H, Davis SC, Jiang SD, Pogue BW, Paulsen KD, Patterson MS. Spectrally resolved bioluminescence optical tomography. Opt Lett. 2006;31(3):365–7.
26. Dehghani H, Davis SC, Pogue BW. Spectrally resolved bioluminescence tomography using the reciprocity approach. Med Phys [Internet]. 2008;35(11):4863–71. http://www.pubmedcentral.nih.gov/articlerender.fcgi?artid=2737244&tool=pmcentrez&rendertype=abstract.
27. He X, Liang J, Wang X, Yu J, Qu X, Wang X, et al. Sparse reconstruction for quantitative bioluminescence tomography based on the incomplete variables truncated conjugate gradient method. Opt Express [Internet]. 2010;18(24):24825–41. https://doi.org/10.1364/oe.18.024825
28. Wooten HO, Green O, Yang M, DeWees T, Kashani R, Olsen J, et al. Quality of intensity modulated radiation therapy treatment plans using a (60)Co magnetic resonance image guidance radiation therapy system. Int J Radiat Oncol Biol Phys [Internet]. 2015;92(4):771–8. http://www.sciencedirect.com/science/article/pii/S0360301615003004.
29. Wooten HO, Rodriguez V, Green O, Kashani R, Santanam L, Tanderup K, et al. Benchmark IMRT evaluation of a Co-60 MRI-guided radiation therapy system. Radiother Oncol [Internet]. 2015;114(3):402–5. https://doi.org/10.1016/j.radonc.2015.01.015.
30. Lagendijk JJW, Raaymakers BW, Raaijmakers AJE, Overweg J, Brown KJ, Kerkhof EM, et al. MRI/linac integration. Radiother Oncol. 2008;86(1):25–9.
31. Schmidt MA, Payne GS. Radiotherapy planning using MRI. Phys Med Biol [Internet]. 2015;60(22):R323–61. http://stacks.iop.org/0031-9155/60/i=22/a=R323?key=crossref.73d6c6caed0c84bc13887e1da0f28de1.
32. Bolcaen J, Descamps B, Deblaere K, Boterberg T, Hallaert G, Van den Broecke C, et al. MRI-guided 3D conformal arc micro-irradiation of a F98 glioblastoma rat model using the Small Animal Radiation Research Platform (SARRP). J Neuro-Oncol. 2014;120(2):257–66.
33. Bryant MJ, Chuah TL, Luff J, Lavin MF, Walker DG. A novel rat model for glioblastoma multiforme using a bioluminescent F98 cell line. J Clin Neurosci. 2008;15(5):545–51.
34. Cherry SR. In vivo molecular and genomic imaging: new challenges for imaging physics. Phys Med Biol. 2004;49(3):R13–48.
35. Gutierrez S, Descamps B, Vanhove C. MRI-only based radiotherapy treatment planning for the rat brain on a Small Animal Radiation Research Platform (SARRP). PLoS One [Internet]. 2015;10(12):e0143821. http://dx.plos.org/10.1371/journal.pone.0143821.
36. Ling CC, Humm J, Larson S, Amols H, Fuks Z, Leibel S, et al. Towards multidimensional radiotherapy (MD-CRT):biological imaging and biological conformality. Int J Radiat Oncol Biol Phys. 2000;47(3):551–60.
37. Wahl RL, Jacene H, Kasamon Y, Lodge MA. From RECIST to PERCIST: evolving considerations for PET response criteria in solid tumors. J Nucl Med [Internet]. 2009;50 Suppl 1(5):122S–50S. http://www.pubmedcentral.nih.gov/articlerender.fcgi?artid=2755245&tool=pmcentrez&rendertype=abstract.
38. Lopci E, Grassi I, Chiti A, Nanni C, Cicoria G, Toschi L, et al. PET radiopharmaceuticals for imaging of tumor hypoxia: a review of the evidence. Am J Nucl Med Mol Imaging [Internet]. 2014;4(4):365–84. http://www.ncbi.nlm.nih.gov/pubmed/24982822.
39. Gregoire V, Chiti A. PET in radiotherapy planning: particularly exquisite test or pending and experimental tool? Radiother Oncol. 2010;96(3):275–6.
40. Chiti A, Kirienko M, Grégoire V. Clinical use of PET-CT data for radiotherapy planning: what are we looking for? Radiother Oncol [Internet]. 2010;96(3):277–9. https://doi.org/10.1016/j.radonc.2010.07.021
41. Delikgoz Soykut E, Ozsahin EM, Yukselen Guney Y, Aytac Arslan S, Derinalp Or O, Altundag MB, et al. The use of PET/CT in radiotherapy planning: contribution of deformable registration. Front Oncol [Internet]. 2013;3(April):33. http://www.pubmedcentral.nih.gov/articlerender.fcgi?artid=3624079&tool=pmcentrez&rendertype=abstract.

42. Schütze C, Bergmann R, Yaromina A, Hessel F, Kotzerke J, Steinbach J, et al. Effect of increase of radiation dose on local control relates to pre-treatment FDG uptake in FaDu tumours in nude mice. Radiother Oncol. 2007;83(3):311–5.
43. Trani D, Yaromina A, Dubois L, Granzier M, Peeters SGJA, Biemans R, et al. Preclinical assessment of efficacy of radiation dose painting based on intratumoral FDG-PET uptake. Clin Cancer Res [Internet]. 2015;21:5511–9. http://clincancerres.aacrjournals.org/cgi/doi/10.1158/1078-0432.CCR-15-0290.
44. España S, Marcinkowski R, Keereman V, Vandenberghe S, Van Holen R. DigiPET: sub-millimeter spatial resolution small-animal PET imaging using thin monolithic scintillators. Phys Med Biol [Internet]. 2014;59(13):3405. http://stacks.iop.org/0031-9155/59/i=13/a=3405.

MIX
Papier aus verantwortungsvollen Quellen
Paper from responsible sources
FSC® C105338

If you have any concerns about our products,
you can contact us on
**ProductSafety@springernature.com**

In case Publisher is established outside the EU,
the EU authorized representative is:
**Springer Nature Customer Service Center GmbH**
**Europaplatz 3, 69115 Heidelberg, Germany**

Printed by Libri Plureos GmbH
in Hamburg, Germany